David Brewster

Life of Sir Isaac Newton

David Brewster

Life of Sir Isaac Newton

ISBN/EAN: 9783741160127

Manufactured in Europe, USA, Canada, Australia, Japa

Cover: Foto ©Andreas Hilbeck / pixelio.de

Manufactured and distributed by brebook publishing software
(www.brebook.com)

David Brewster

Life of Sir Isaac Newton

LIFE

OF

SIR ISAAC NEWTON.

THE LIFE

OF

SIR ISAAC NEWTON.

BY

SIR DAVID BREWSTER,

K.H. D.C.L. F.R.S.

A NEW EDITION REVISED

BY

W. T. LYNN, B.A. F.R.A.S.

OF THE ROYAL OBSERVATORY GREENWICH

The Birth Place of Newton

LONDON:

THE

LIFE

OF

SIR ISAAC NEWTON,

BY

SIR DAVID BREWSTER,

K.H., LL.D., F.R.S., V.P.R.S.E., ETC.

A NEW EDITION.

REVISED AND EDITED

BY

W. T. LYNN, B.A., F.R.A.S.,

OF THE

ROYAL OBSERVATORY, GREENWICH.

———

LONDON:
WILLIAM TEGG & CO.,
PANCRAS LANE, QUEEN STREET, CHEAPSIDE.
1875.

ADVERTISEMENT.

It will be in the recollection of many of our readers
that Sir David Brewster's long and useful life closed on
the 10th of February, 1868. That one so distinguished
in natural philosophy should write the biography of Sir
Isaac Newton, who by common consent stands at the
head of all natural philosophers of all ages, must seem
exceedingly appropriate; and it is generally allowed that
he accomplished his difficult task in a manner worthy of
the subject. In passing this new Edition of it through
the press, I have carefully preserved the author's text as
it stands, and have simply affixed such notes as in the
present day appear to me to be desirable. The text
followed is principally that of 1831, but it has been
collated throughout with that of the larger edition
published in two volumes in 1855, and thus every
possible accuracy, according to the author's latest researches, secured.

W. T. L.

CONTENTS.

CHAPTER IV.

CHAPTER V.

CHAPTER VI.

CHAPTER VII.

CHAPTER VIII.

CHAPTER IX.

CHAPTER X.

CHAPTER XI.

CHAPTER XII.

CHAPTER XIX.

LIFE

OF

SIR ISAAC NEWTON.

CHAPTER I.

The Pre-eminence of Sir Isaac Newton's Reputation—The Interest attached to the Study of his Life and Writings— His Birth and Parentage—His Early Education—Is sent to Grantham School—His Early Attachment to Mechanical Pursuits—His Windmill—His Water-Clock—His Self-moving Cart—His Sun-Dials—His Preparation for the University.

THE name of Sir Isaac Newton has by general consent been placed at the head of those great men who have been the ornaments of their species. However imposing be the attributes with which time has invested the sages and the heroes of antiquity, the brightness of their fame has been eclipsed by the splendour of his reputation; and neither the partiality of rival nations, nor the vanity of a presumptuous age, has ventured to dispute the ascendancy of his genius. The philosopher,* indeed, to whom posterity will probably assign the place next to Newton, has characterized the *Principia* as pre-eminent above all the productions of human intellect, and has thus divested of extravagance the contemporary encomium upon its author,—

> Nec fas est proprius mortali attingere divos.
>
> So near the gods—man cannot nearer go.
>
> HALLEY.

---

* The Marquis La Place.—See *Système du Monde*, p. 336.

The biography of an individual so highly renowned
cannot fail to excite a general interest. Though his
course may have lain in the vale of private life, and may
have been unmarked with those dramatic events which
throw a lustre even round perishable names, yet the in-
quiring spirit will explore the history of a mind so richly
endowed,—will study its intellectual and moral phases,—
and will seek the shelter of its authority on those great
questions which reason has abandoned to faith and hope.

If the conduct and opinions of men of ordinary talent
are recorded for our instruction, how interesting must it
be to follow the most exalted genius through the incidents
of common life; to mark the steps by which he attained
his lofty pre-eminence; to see how he performs the
functions of the social and the domestic compact; how
he exercises his lofty powers of invention and discovery;
how he comports himself in the arena of intellectual strife;
and in what sentiments, and with what aspirations, he
quits the world which he has adorned!

In almost all these bearings, the life and writings of
Sir Isaac Newton abound with the richest counsel. Here
the philosopher will learn the art by which alone he can
acquire an immortal name. The moralist will trace the
lineaments of a character adjusted to all the symmetry of
which our imperfect nature is susceptible; and the
Christian will contemplate with delight the high priest of
science quitting the study of the material universe,—the
scene of his intellectual triumphs,—to investigate with
humility and patience the mysteries of his faith.

Sir Isaac Newton was born at Woolsthorpe, a hamlet
in the parish of Colsterworth in Lincolnshire, about eight
miles south of Grantham,* on the 25th of December O. S.

* In these days, it will be perhaps useful to mention that the nearest
railroad station is Great Ponton (on the main Great Northern Line from
Grantham to Peterborough), which is about four miles from Woolsthorpe.

1642, exactly one year after Galileo died, and was baptized at Colsterworth, on the 1st of January, 1642-3.   His father, Mr. Isaac Newton, died at the early age of thirty-six, a little more than a year after the death of his father, Robert Newton, and only a few months after his marriage to Harriet Ayscough, daughter of James Ayscough, of Market Overton, in Rutlandshire.   This lady was accordingly left in a state of pregnancy, and appears to have given a premature birth to her only and posthumous child. The helpless infant thus ushered into the world, was of such an extremely diminutive size, * and seemed of so perishable a frame, that two women, who were sent to Lady Pakenham's at North Witham, to bring some medicine to strengthen him, did not expect to find him alive on their return.   Providence, however, had otherwise decreed: and that frail tenement, which seemed scarcely able to imprison its immortal mind, was destined to enjoy a vigorous maturity, and to survive even the average term of human existence.   The estate of Woolsthorpe, in the manor-house of which this remarkable birth took place, had been more than an hundred years in the possession of the family, who came originally from Newton in Lancashire, but who had, previous to the purchase of Woolsthorpe, settled at Westby, in the county of Lincoln.   The manor-house, of which we have given an engraving, is situated in a beautiful little valley, . remarkable for its copious wells of pure spring water, on the left side of the river Witham, which has its origin in the neighbourhood, and commands an agreeable prospect to the east towards Colsterworth.   The manor of

That hamlet is only about half a mile from Colsterworth, and must not be confounded with a larger village, also called Woolsthorpe, in the same county, and also in the "parts" of Kesteven.—EDITOR.

* Sir Isaac Newton told Mr. Conduit, that he had often heard his mother say, that when he was born he was so little that they might have put him into a quart mug.

Woolsthorpe was worth only £30 per annum; but Mrs. Newton possessed another small estate at Sewstern,* which raised the annual value of their property to about £80; and it is probable that the cultivation of the little farm† on which she resided, somewhat enlarged the limited income upon which she had to support herself and educate her child.

For three years Mrs. Newton continued to watch over her tender charge with parental anxiety; but in consequence of her marriage to the Rev. Barnabas Smith, rector of North Witham, about a mile south of Woolsthorpe, she left him under the care of her own mother. At the usual age he was sent to two day-schools at Skillington and Stoke, where he acquired the education which such seminaries afforded; but when he reached his twelfth year he went to the public school at Grantham,‡ taught by Mr. Stokes, and was boarded at the house of Mr. Clark, an apothecary in that town. According to information which Sir Isaac himself gave to Mr. Conduit, he seems to have been very inattentive to his studies, and very low in the school. The boy, however, who was above him, having one day given him a severe kick upon his stomach, from which he suffered great pain, Isaac laboured incessantly till he got above him in the school, and from that time he continued to rise till he was the head boy. From the habits of application which this incident had led him to form, the peculiar character of his mind was speedily displayed. During the hours of play, when the other boys were occupied with their amusements, his mind was en-

* In Leicestershire, and about three miles south-east of Woolsthorpe.

† It is now in the occupation of a farmer named Woollerton. At a recent visit, we were politely shown over the house by Mrs. Woollerton.—EDITOR.

‡ This is now in the able hands of Mr. Beasley, of St. John's College, Cambridge, author of a well-known book on Trigonometry.—EDITOR.

grossed with mechanical contrivances, either in imitation
of something which he had seen, or in execution of some
original conception of his own.   For this purpose he provided himself with little saws, hatchets, hammers, and all
sorts of tools, which he acquired the art of using with
singular dexterity.   The principal pieces of mechanism
which he thus constructed were a wind-mill, a water-
clock, and a carriage put in motion by the person who sat
in it.   When a wind-mill was erecting near Grantham,
on the road to Gunnerby, Isaac frequently attended the
operations of the workmen, and acquired such a thorough
knowledge of the machinery, that he completed a working
model of it, which excited universal admiration.   This
model was frequently placed at the top of the house in
which he lodged at Grantham, and was put in motion by
the action of the wind upon its sails.   Not content with
this exact imitation of the original machine, he conceived
the idea of driving it by animal power ; and for this purpose he enclosed in it a mouse, which he called the miller,
and which, by acting upon a sort of tread-wheel, gave
motion to the machine.   According to some accounts, the
mouse was made to advance by pulling a string attached
to its tail, while others allege that the power of the little
agent was called forth by its unavailing attempts to reach
a portion of corn placed above the wheel.

His water-clock was formed out of a box which he had
solicited from Mrs. Clark's brother.   It was about four
feet high, and of a proportional breadth, somewhat like a
common house-clock.   The index of the dial-plate was
turned by a piece of wood, which either fell or rose by the
action of dropping water.   As it stood in his own bedroom,
he supplied it every morning with the requisite quantity
of water, and it was used as a clock by Mr. Clark's family,
and remained in the house long after its inventor had

quitted Grantham.* His mechanical carriage was a vehicle with four wheels, which was put in motion with a handle wrought by the person who sat in it; but, like Merlin's chair, it seems to have been used only on the smooth surface of a floor, and not fitted to overcome the inequalities of a road. Although Newton was at this time "a sober, silent, thinking lad," who scarcely ever joined in the ordinary games of his school-fellows, yet he took great pleasure in providing them with amusements of a scientific character. He introduced into the school the flying of paper kites; and he is said to have been at great pains in determining their best forms and proportions, and in ascertaining the position and number of the points by which the string should be attached. He made also paper lanthorns, by the light of which he went to school in the winter mornings, and he frequently attached these lanthorns to the tails of his kites in a dark night, so as to inspire the country people with the belief that they were comets.

In the house where he lodged there were some female inmates, in whose company he appears to have taken much pleasure. One of these, a Miss Storey, sister to Dr. Storey, a physician at Buckminster, near Colsterworth, was two or three years younger than Newton, and to great personal attractions she seems to have added more than the usual allotment of female talent. The society of this young lady and her companions was always preferred to that of his own school-fellows, and it was one of his most agreeable occupations to construct for them little tables and cup-

* "I remember once," says Dr. Stukely, "when I was deputy to Dr. Halley, Secretary at the Royal Society, Sir Isaac talked of these kind of instruments. That he observed the chief inconvenience in them was, that the hole through which the water is transmitted being necessarily very small, was subject to be furred up by impurities in the water, as those made with sand will wear bigger, which at length causes an inequality in time."—Stukely's Letter to Dr. Mead.—Turner's *Collections*, p. 177.

boards, and other utensils for holding their dolls and their trinkets. He had lived nearly six years in the same house with Miss Storey, and there is reason to believe that their youthful friendship gradually rose to a higher passion; but the smallness of her portion and the inadequacy of his own fortune appear to have prevented the consummation of their happiness. Miss Storey was afterwards twice married, and under the name of Mrs. Vincent, Dr. Stukely visited her at Grantham in 1727, at the age of eighty-two, and obtained from her many particulars respecting the early history of our author. Newton's esteem for her continued unabated during his life. He regularly visited her when he went to Lincolnshire, and never failed to relieve her from little pecuniary difficulties which seem to have beset her family.

Among the early passions of Newton we must recount his love of drawing, and even of writing verses. His own room was furnished with pictures drawn, coloured, and framed by himself, sometimes from copies, but often from life.* Among these were portraits of Dr. Donne, Mr. Stokes, the master of Grantham School, and King Charles I., under whose picture were the following verses :—

A secret art my soul requires to try,
If prayers can give me what the wars deny.
Three crowns distinguished here, in order do
Present their objects to my knowing view.
Earth's crown, thus at my feet I can disdain,
Which heavy is, and at the best but vain.
But now a crown of thorns I gladly greet;
Sharp is this crown, but not so sharp as sweet;
The crown of glory that I yonder see
Is full of bliss and of eternity.

* Mr. Clark informed Dr. Stukely that the walls of the room in which Sir Isaac lodged were covered with charcoal drawings of birds, beasts, men, ships, and mathematical figures, all of which were very well designed.

These verses were repeated to Dr. Stukely by Mrs. Vincent, who believed them to be written by Sir Isaac, a circumstance which is the more probable, as he himself assured Mr. Conduit, with some expression of pleasure, that he "excelled in making verses," although he had been heard to express a contempt for poetical composition.

But while the mind of our young philosopher was principally occupied with the pursuits which we have now detailed, it was not inattentive to the movements of the celestial bodies on which he was destined to throw such a brilliant light. The imperfections of his water-clock had probably directed his thoughts to the more accurate measure of time which the motion of the sun afforded. In the yard of the house where he lived, he traced the varying movements of that luminary upon the walls and roofs of the buildings, and by means of fixed pins he had marked out the hourly and half-hourly subdivisions. One of these dials, which went by the name of *Isaac's dial*, and was often referred to by the country people for the hour of the day, appears to have been drawn solely from the observations of several years; but we are not informed whether all the dials which he drew on the walls of his house at Woolsthorpe, and which existed after his death, were of the same description, or were projected from his knowledge of the doctrine of the sphere.

Upon the death of the Rev. Mr. Smith in the year 1656, his widow left the rectory of North Witham, and took up her residence at Woolsthorpe, along with her three children, Mary, Benjamin, and Hannah Smith. Newton had now attained the fifteenth year of his age, and had made great progress in his studies; and as he was thought capable of being useful in the management of the farm and country business at Woolsthorpe, his mother, chiefly from a motive of economy, recalled him from the school at Grantham. In order to accustom him

to the arts of selling and buying, two of the most important branches of rural labour, he was frequently sent on
Saturday to Grantham market, to dispose of grain and
other articles of farm produce, and to purchase such
necessaries as the family required. As he had yet acquired no experience, an old trustworthy servant generally accompanied him on these errands. The Inn which
they patronized was the Saracen's Head at West Gate;
but no sooner had they put up their horses than our young
philosopher deserted his commercial concerns, and betook
himself to his former lodging in the apothecary's garret,
where a number of Mr. Clark's old books afforded him
abundance of entertainment till his aged guardian had
executed the family commissions, and announced to him
the necessity of returning. At other times he deserted
his duties at an earlier stage, and intrenched himself
under a hedge by the way side, where he continued his
studies till the servant returned from Grantham. The
more immediate affairs of the farm were not more pros-
perous under his management than would have been his
marketings at Grantham. The perusal of a book, the
execution of a model, or the superintendence of a water-
wheel of his own construction, whirling the glittering
spray from some neighbouring stream, absorbed all his
thoughts; whilst the sheep were going astray, and the
cattle were devouring or treading down the corn.

Mrs. Smith was soon convinced from experience that
her son was not destined to cultivate the soil; and as his
passion for study, and his dislike for every other occupation, increased with his years, she wisely resolved to give
him all the advantages which education could confer. He
was accordingly sent back to Grantham school, where he
continued for some months in busy preparation for his
academical studies. His uncle, the Rev. W. Ayscough,
who was rector of Burton Coggles, about three miles east

of Woolsthorpe, and who had himself studied at Trinity College, recommended to his nephew to enter that society, and it was accordingly determined that he should proceed to Cambridge at the approaching term.*

* "One of his uncles," says M. Biot, "having one day found him under a hedge with a book in his hand, and entirely absorbed in meditation, took it from him, and found that he was occupied in the solution of a mathematical problem. Struck with finding so serious and so active a disposition at so early an age, he urged his mother no longer to thwart him, and to send him back to Grantham to continue his studies." I have omitted this anecdote in the text, as I cannot find it in Turner's Collections, from which M. Biot derived his details of Newton's infancy, nor in any other work.

CHAPTER II.

Newton enters Trinity College, Cambridge — Origin of his Propensity for Mathematics — He studies the Geometry of Descartes unassisted — Purchases a Prism — Revises Dr. Barrow's Optical Lectures — Dr. Barrow's Opinion respecting Colours — Takes his Degrees — Is appointed a Fellow of Trinity College — Succeeds Dr. Barrow in the Lucasian Chair of Mathematics.

To a young mind thirsting for knowledge, and ambitious of the distinction which it brings, the transition from a village school to an university like that of Cambridge, — from the absolute solitude of thought to the society of men imbued with all the literature and science of the age, — must be one of eventful interest. To Newton it was a source of peculiar excitement. The history of science affords many examples where the young aspirant had been early initiated into her mysteries, and had even exercised his powers of invention and discovery before he was admitted within the walls of a college; but he who was to give philosophy her laws did not exhibit such early talent;—no friendly counsel regulated his youthful studies, and no work of scientific eminence seems to have guided him in his course. In yielding to the impulse of his mechanical genius, his mind obeyed the laws of its own natural expansion; and, following the line of least resistance, it was thus drawn aside from the strongholds with which it was destined to grapple.

When Newton, therefore, arrived at Trinity College,

he brought with him a more slender portion of science than falls to the lot of ordinary scholars; but this state of his acquirements was perhaps not unfavourable to the developement of his powers. Unexhausted by premature growth, and invigorated by healthful repose, his mind was the better fitted to make those vigorous and rapid shoots, which soon covered with foliage and with fruit the genial soil to which it had been transferred.

Cambridge was consequently the real birth-place of Newton's genius. Her teachers fostered his earliest studies; her institutions sustained his mightiest efforts; and within her precincts were all his discoveries made and perfected. When he was called to higher official functions, his disciples kept up the pre-eminence of their master's philosophy; and their successors have maintained this seat of learning in the fulness of its glory, and rendered it the most distinguished among the universities of Europe.

It was on the 5th of June, 1660, in the 18th year of his age, that Newton was admitted into Trinity College, Cambridge, during the same year that Dr. Barrow was elected Professor of Greek in the university. His attention was first turned to the study of mathematics by a desire to inquire into the truth of judicial astrology; and he is said to have discovered the folly of that study by erecting a figure with the aid of one or two of the problems of Euclid. The propositions contained in this ancient system of geometry, he regarded as self-evident truths; and without any preliminary study he made himself master of Descartes's Geometry by his genius and patient application. This neglect of the elementary truths of geometry he afterwards regarded as a mistake in his mathematical studies; and he expressed to Dr. Pemberton his regret "that he had applied himself to the works of Descartes and other algebraic writers, before he had considered the elements of Euclid with that attention which so excellent a writer deserved."* Dr. Wallis's Arithmetic of Infinites, Sanderson's Logic, and the Optics of Kepler were among the books which he had studied with care. On the works he wrote comments during their perusal; and so great was his progress, that he is reported to have found himself more deeply versed in some branches of knowledge than the tutor who directed his studies.

Neither history nor tradition has handed down to us any particular account of his progress during the first three years that he spent at Cambridge. It appears from a statement of his expenses, that in 1664 he purchased a prism, for the purpose, as has been said, of examining Descartes's theory of colours; and it is stated by Mr. Conduit, that he soon established his own views on the subject, and detected the errors in those of the French philosopher. This, however, does not seem to have been the case. Had he discovered the composition of Light in 1664 or 1665, it is not likely that he would have withheld it not only from the Royal Society, but from his own friends at Cambridge, till the year 1671. His friend and tutor, Dr. Barrow, was made Lucasian Professor of Mathematics in 1663, and the optical lectures which he afterwards delivered were published in 1669. In the preface of this work he acknowledges his obligations to his colleague, Mr. Isaac Newton,† for having revised the MSS. and corrected several oversights, and made some important suggestions. In the twelfth lecture there are some observations on the nature and origin of colours, which are so erroneous and unphilosophical, that Newton could not have permitted his friend to publish them, had he been then in possession of the true theory. According to Barrow, who introduces the subject of colours as an

* Pemberton's *View of Sir Isaac Newton's Philosophy.* Pref.

† Peregregiæ vir indolis ac insignis peritiæ.—*Epist. ad Lect.*

unusual digression:—*White* is that which discharges a copious light equally clear in every direction—*Black* is that which does not emit light at all, or which does it very sparingly—*Red* is that which emits a light more clear than usual, but interrupted by shady interstices—*Blue* is that which discharges a rarefied light, as in bodies which consist of white and black particles arranged alternately—*Green* is nearly allied to blue—*Yellow* is a mixture of much white and a little red—and *Purple* consists of a great deal of blue mixed with a small portion of red. The blue colour of the sea arises from the whiteness of the salt which it contains, mixed with the blackness of the pure water in which the salt is dissolved; and the blueness of the shadows of bodies, seen at the same time by candle and day light, arises from the whiteness of the paper mixed with the faint light or blackness of the twilight. These opinions savour so little of genuine philosophy that they must have attracted the observation of Newton; and had he discovered at that time that white was a mixture of all the colours, and black a privation of them all, he could not have permitted the absurd speculations of his master to pass uncorrected.

That Newton had not distinguished himself by any positive discovery so early as 1664 or 1665, may be inferred also from the circumstances which attended the competition for the law fellowship of Trinity College. The candidates for this appointment were himself and Mr. Robert Uvedale; and Dr. Barrow, then Master of Trinity, having found them perfectly equal in their attainments, conferred the fellowship on Mr. Uvedale as the senior candidate.

In the books of the University, Newton is recorded as having been admitted sub-sizer in 1661. He became a scholar in 1664. In 1665 he took his degree of Bachelor of Arts, and in 1666, in consequence of the breaking out

of the plague, he retired to Woolsthorpe. In 1667 he
was made Junior Fellow. In 1668 he took his degree of
Master of Arts, and in the same year he was appointed to
a Senior Fellowship. In 1669, when Dr. Barrow had
resolved to devote his attention to theology, he resigned
the Lucasian Professorship of Mathematics in favour of
Newton, who may now be considered as having entered
upon that brilliant career of discovery, the history of
which will form the subject of some of the following
chapters.

CHAPTER III.

*Newton occupied in grinding hyperbolical Lenses—His first
Experiments with the Prism made in 1666—He discovers
the Composition of White Light, and the different Refrangibility of the Rays which compose it—Abandons his attempts
to improve Refracting Telescopes, and resolves to attempt
the Construction of Reflecting ones—He quits Cambridge on
Account of the Plague—Constructs two Reflecting Telescopes
in 1668, the first ever executed—One of them examined by
the Royal Society, and shown to the King—He constructs
a Telescope with Glass Specula—Recent History of the
Reflecting Telescope—Airy's Glass Specula—Hadley's
Reflecting Telescopes—Short's—Herschel's—Supposed
Decline of Science in England.*

THE appointment of Newton to the Lucasian Chair at
Cambridge seems to have been coëval with his grandest
discoveries. The first of these, of which the date is well
authenticated, is that of the different refrangibility of the
rays of light, which he established in 1666. The germ of
the doctrine of universal gravitation seems to have presented itself to him in the same year, or at least in 1667 ;
and "in the year 1666, or before"* he was in possession
of his method of fluxions ; and he had brought it to such
a state in the beginning of 1669, that he permitted Dr.
Barrow to communicate it to Mr. Collins on the 20th
of June in that year.

Although we have already mentioned, on the authority

* See Newton's Letter to the Abbé Conti, dated February 26th, 1715-16,
in the *Additamenta Comm. Epistolici.*

of a written memorandum of Newton himself, that he purchased a prism at Cambridge in 1664, yet he does not appear to have made any use of it, as he informs us that it was in 1666 that he procured a triangular glass prism, to try therewith the celebrated phenomena of colours."[*] During that year he had applied himself to the grinding of " optick glasses, of other figures than spherical ;" and having, no doubt, experienced the impracticability of executing such lenses, the idea of examining the phenomena of colour was one of those sagacious and fortunate impulses which more than once led him to discovery. Descartes, in his *Dioptrice*, published in 1629, and more recently James Gregory, in his *Optica Promota*, published in 1669, had shown that parallel and diverging rays could be reflected or refracted, with mathematical accuracy, to a point or focus, by giving the surface a parabolic, an elliptical, or a hyperbolic form, or some other form not spherical. Descartes had even invented and described machines by which lenses of these shapes could be ground and polished; and the perfection of the refracting telescope was supposed to depend on the degree of accuracy with which they could be executed.

In attempting to grind glasses that were not spherical, Newton seems to have conjectured that the defects of lenses, and consequently of refracting telescopes, might arise from some other cause than the imperfect convergency of rays to a single point; and this conjecture was happily realized in those fine discoveries of which we shall now endeavour to give some account.

When Newton began this inquiry, philosophers of the highest genius were directing all the energies of their minds to the subject of light, and to the improvement of the refracting telescope. James Gregory of Aberdeen

---

* Newton *Opera*, tom. iv., p. 205. Letter to Oldenburg.

had invented his reflecting telescope. Descartes had explained the theory, and exerted himself in perfecting the constructing of the common refracting telescope; and Huygens had not only executed the magnificent instruments by which he discovered the ring and the satellites of Saturn, but had begun those splendid researches respecting the nature of light, and the phenomena of double refraction, which have led his successors to such brilliant discoveries. Newton, therefore, arose when the science of light was ready for some great accession, and at the precise time when he was required to propagate the impulse which it had received from his illustrious predecessors.

The ignorance which then prevailed respecting the nature and origin of colours, is sufficiently apparent from the account we have already given of Dr. Barrow's speculations on this subject. It was always supposed that light of every colour was equally refracted or bent out of its direction when it passed through any lens or prism, or other refracting medium; and though the exhibition of colours by the prism had been often made previous to the time of Newton, yet no philosopher seems to have attempted to analyse the phenomena.

When he had procured his triangular glass prism, a section of which is shown at ABC, Fig. 1, he made a

Fig. 1.

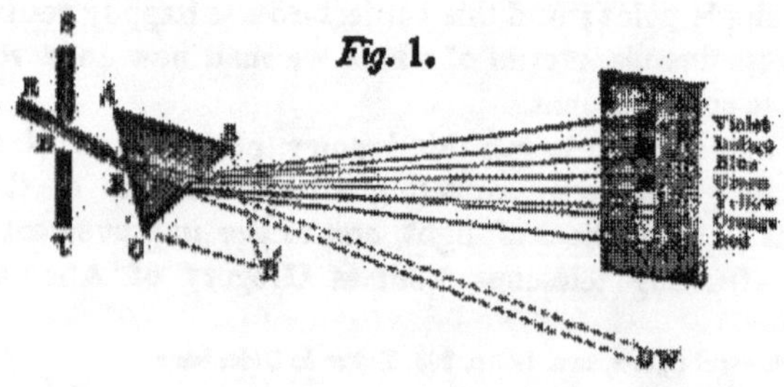

hole H in one of his window shutters, SHT; and, having darkened his chamber, he let in a convenient quantity of the sun's light RR, which, passing through the prism ABC, was so refracted as to exhibit all the different colours on the wall at MN, forming an image about five times as long as it was broad. "It was at first," says our author, " a very pleasing divertisement to view the vivid and intense colours produced thereby ; " but this pleasure was immediately succeeded by surprise at various circumstances which he had not expected. According to the received laws of refraction, he expected the image MN to be circular, like the white image at W, which the sunbeam RR had formed on the wall previous to the interposition of the prism ; but when he found it to be no less than five times larger than its breadth, it excited in him a more than ordinary curiosity to examine from whence it might proceed. He could scarcely think that the various thickness of the glass, or the termination with shadow or darkness, could have any influence on light to produce such an effect; yet he thought it not amiss first to examine those circumstances, and so find what would happen by transmitting light through parts of the glass of divers thicknesses, or through holes in the window of divers bignesses, or by setting the prism without, (on the other side of ST,) so that the light might pass through it and be refracted before it was terminated by the hole; but he found none of these circumstances material. " The fashion of the colours was in all those cases the same."

Newton next suspected that some unevenness in the glass, or other accidental irregularity, might cause the dilatation of the colours. In order to try this he took another prism BCB', and placed it in such a manner that the light RRW passing through them both might be refracted contrary ways, and thus returned by BCB' into that course RRW from which the prism ADC had di-

verted it; for by this means he thought the regular effects
of the prism ABC would be destroyed by the prism BCB',
and the irregular ones more augmented by the multiplicity
of refractions. The result was, that the light which was
diffused by the first prism ABC into an oblong form, was
reduced by the second prism BCB' into a circular one W,
with as much regularity as when it did not pass through
them at all; so that whatever was the cause of the length
of the image MN, it did not arise from any irregularity
in the prism.

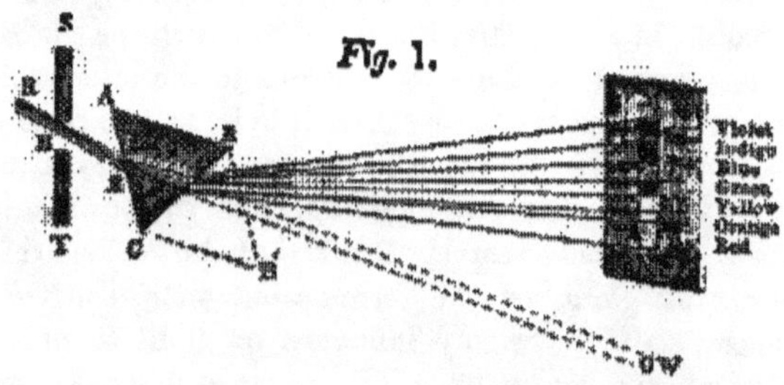

Our author next proceeded to examine more critically
what might be effected by the difference of the incidence
of the rays proceeding from different parts of the sun's
disc; but by taking accurate measures of the lines and
angles, he found that the angle of the emergent rays
should be 31 minutes, equal to the sun's diameter, whereas
the real angle subtended by MN at the hole H was 2° 49'.
But as this computation was founded on the hypothesis,
that the sine of the angle of incidence was proportional
to the sine of the angle of refraction, which, from his own
experience, he could not imagine to be so erroneous as to
make that angle but 31' which was in reality 2° 49', yet
"his curiosity caused him again to take up his prism"
ABC, and having turned it round in both directions, so
as to make the rays RR fall both with greater and with

less obliquity upon the face AC, he found that the colours on the wall did not sensibly change their place; and hence he obtained a decided proof that they could not be occasioned by a difference in the incidence of the light radiating from different parts of the sun's disc.

Newton then began to suspect that the rays, after passing through the prism, might move in curve lines, and, in proportion to the different degrees of curvature, might tend to different parts of the wall; and this suspicion was strengthened by the recollection, that he had often seen a tennis-ball, struck with an oblique racket, describe such a curve line. In this case a circular and a progressive motion is communicated to the ball by the stroke, and in consequence of this the direction of its motion was curvilineal; so that if the rays of light were globular bodies, they might acquire a circulating motion by their oblique passage out of one medium into another, and thus move like the tennis-ball in a curve line. Notwithstanding, however, " this plausible ground of suspicion," he could discover no such curvature in their direction; and, what was enough for his purpose, he observed that the difference between the length MN of the image, and the diameter of the hole H, was proportional to their distance HM, which could not have happened had the rays moved in curvilineal paths.

These different hypotheses, or suspicions, as Newton calls them, being thus gradually removed, he was at length led to an experiment which determined beyond a doubt the true cause of the elongation of the coloured image. Having taken a board with a small hole in it, he placed it behind the face BC of the prism, and close to it, so that he could transmit through the hole any one of the colours in MN, and keep back all the rest. When the hole, for example, was near C, no other light but the red fell upon the wall at N. He then placed behind N

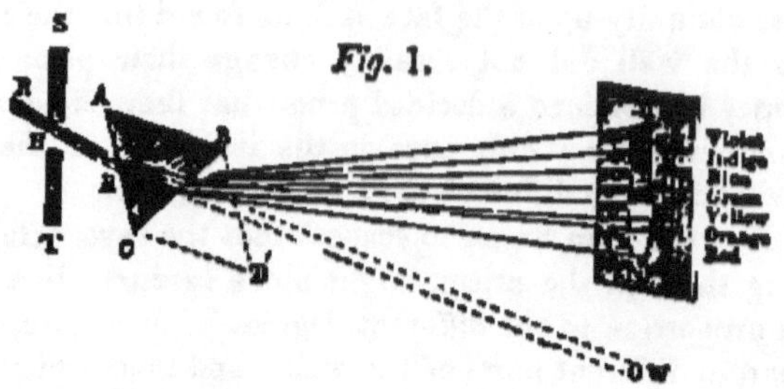

another board with a hole in it, and behind this board he placed another prism, so as to receive the red light at N, which passed through this hole in the second board. He then turned round the first prism ABC, so as to make all the colours pass in succession through these two holes, and he marked their places on the wall. From the variation of these places, he saw that the *red* rays at N were less refracted by the second prism than the *orange* rays, the *orange* less than the yellow, and so on, the *violet* being more refracted than all the rest.

Hence he drew the grand conclusion, *that light was not homogeneous, but consisted of rays, some of which were more refrangible than others.*

Having established this important truth, Newton immediately perceived that a lens which refracts light exactly like a prism, must also refract the differently coloured rays with different degrees of force, bringing the violet rays to a focus nearer the glass than the red rays. This is shown in Fig. 2, where LL is a convex lens and SL, SL rays of the sun falling upon it in parallel directions. The *violet* rays existing in the white light SL being more refrangible than the rest, will be more refracted or bent, and will meet at V, forming there a *violet* image of the sun. In like manner the *yellow* rays will form an image of the sun at Y, and so on; the *red* rays, which are the

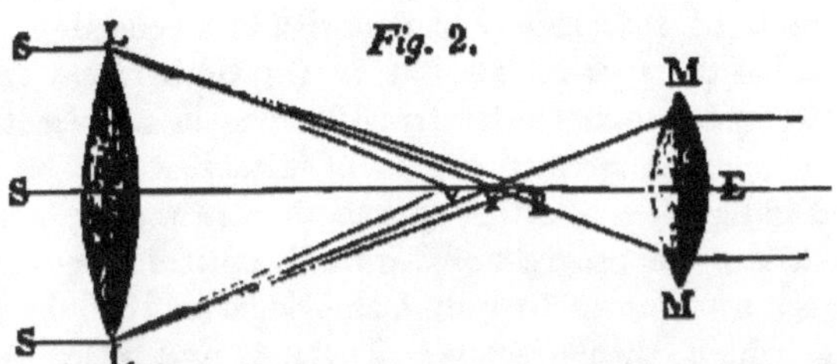

least refrangible, being brought to a focus at R, and there forming a red image of the sun.

Hence, if we suppose LL to be the object-glass of a telescope directed to the sun, and MM an eye-glass, through which the eye at E sees magnified the image or picture of the sun formed by LL, it cannot see distinctly all the different images between R and V. If it is adjusted so as to see distinctly the *yellow* image at Y, as it is in the figure, it will not see distinctly either the *red* or *violet* images, nor indeed any of them but the *yellow* one. There will consequently be a distinct yellow image, with indistinct images of all the other colours, producing great confusion and indistinctness of vision. As soon as Newton saw the bearing of his discovery upon the imperfection of refracting telescopes, he abandoned his attempts to improve them, and took into consideration the principle of reflexion; and as he found that rays of all colours were reflected regularly, so that the angle of reflexion was equal to the angle of incidence, he concluded that, upon this principle, *optical instruments might be brought to any degree of perfection imaginable*, provided a reflecting substance could be found which could polish as finely as glass, and reflect as much light as glass transmits, and provided a method of communicating to it a parabolic figure could be obtained. These difficulties, however, appeared to him very great, and he even thought them insuperable, when he considered, that, as any irregularity in a reflecting surface makes the rays deviate five or six times more from their true path than similar irregularities in a refracting surface, a much greater degree of nicety would be required in figuring reflecting specula than refracting lenses.

Such was the progress of Newton's optical discoveries when he was forced to quit Cambridge in 1666 by the plague which then desolated England, and more than two years elapsed before he proceeded any farther. In 1666 he resumed the inquiry, and having thought of a delicate method of polishing, proper for metals, by which, as he conceived, " the figure would be corrected to the last," he began to put this method to the test of experiment. At this time he was acquainted with the proposal of Mr. James Gregory, contained in his *Optica Promota,* to construct a reflecting telescope with two concave specula, the largest of which had a hole in the middle of the larger speculum, to transmit the light to an eye-glass; * but he conceived that it would be an improvement of this instrument to place the eye-glass at the side of the tube, and to reflect the rays to it by an oval plane speculum. One of those instruments he actually executed with his own hands; and he gave an account of it in a letter to a

* M. Biot, in his Life of Newton, has stated, that Newton was preceded in the invention of the reflecting telescope by Gregory, *but probably without knowing it.* It is quite certain, however, that Newton was acquainted with Gregory's invention, as appears from the following avowal of it. " When I first applied myself to try the effects of reflexion, Mr. Gregory's *Optica Promota* (printed in the year 1663) having fallen into my hands, where there is an instrument described with a hole in the midst of the object-glass, to transmit the light to an eye-glass placed behind it, I had thence an occasion of considering that sort of construction, and found their disadvantages so great, that I saw it necessary, before I attempted any thing in the practice, to alter the design of them, and place the eye-glass at the side of the tube rather than at the middle."—Letter to Oldenburg, May 4th, 1672.

friend, dated February 23rd, 1668-9, a letter which is also remarkable for containing the first allusion to his discoveries respecting colours. Previous to this he was in correspondence on the subject with Mr. Ent, afterwards Sir George Ent, one of the original council of the Royal Society, an eminent medical writer of his day, and President of the College of Physicians. In a letter to Mr. Ent he had promised an account of his telescope to their mutual friend, and the letter to which we now allude contained the fulfilment of that promise. The telescope was six inches long. It bore an aperture in the large speculum something more than an inch, and as the eye-glass was a plano-convex lens, whose focal length was one-sixth or one-seventh of an inch, it magnified about forty times, which, as Newton remarks, was more than any six foot tube (meaning refracting telescope) could do with distinctness. On account of the badness of the materials, however, and the want of a good polish, it represented objects less distinct than a six-feet tube, though he still thought it would be equal to a three or four feet tube directed to common objects. He had seen through it Jupiter distinctly with his four satellites, and also the horns or moon-like phases of Venus, though this last phenomenon required some niceness in adjusting the instrument.

Although Newton considered this little instrument as in itself contemptible, yet he regarded it as an " epitome of what might be done," and he expressed his conviction that a six feet telescope might be made after this method, which would perform as well as a sixty or a hundred feet telescope made in the common way ; and that if a common refracting telescope could be made of the " purest glass exquisitely polished, with the best figure that any geometrician (Descartes, &c.) hath or can design," it would scarcely perform better than a common telescope. This, he adds, may seem a paradoxical assertion; yet, he continues, " it is the necessary consequence of some experiments which I have made concerning the nature of light."

The telescope now described possesses a very peculiar interest, as being the first reflecting one which was ever executed and directed to the heavens. James Gregory, indeed, had attempted, in 1664 or 1665, to construct his instrument. He employed Messrs. Rives and Cox, who were celebrated glass-grinders of that time, to execute a concave speculum of six feet radius, and likewise a small one; but as they had failed in polishing the large one, and as Mr. Gregory was on the eve of going abroad, he troubled himself no farther about the experiment, and the tube of the telescope was never made. Some time afterwards, indeed, he " made some trials both with a little concave and convex speculum," but, " possessed with the fancy of the defective figure," he " would not be at the pains to fix every thing in its due distance."

Such were the earliest attempts to construct the reflecting telescope, that noble instrument which has since effected such splendid discoveries in astronomy. Looking back from the present state of practical science upon its progressive advance, how great is the contrast between the loose specula of Gregory and the fine Gregorian telescopes of Hadley, Short, and Veitch,—between the humble six inch tube of Newton and the gigantic instruments of Herschel and Lord Rosse! The latter is indeed one of the most wonderful combinations of art and science which the world has yet seen. We have in the morning walked, again and again, and ever with new delight, along its mystic tube; and at midnight, with its distinguished architect, pondered over the marvellous sights which it discloses—the satellites and belts and rings of Saturn— the old and new ring, which is advancing with its crest of waters to the body of the planet—the rocks and

mountains and valleys, and extinct volcanoes of the Moon
—the crescent of Venus with its mountainous outline—
the systems of double and triple stars, the nebulæ and
starry clusters of every variety of shape—and those spiral
nebular formations which baffle human comprehension,
and constitute the greatest achievement in modern discovery.

The success of his first experiment inspired Newton
with fresh zeal; and though his mind was now occupied
with his optical discoveries, with the elements of his method
of fluxions, and with the expanding germ of his theory of
universal gravitation, yet with all the ardour of youth he
applied himself to the laborious operation of executing
another reflecting telescope with his own hands. This instrument, which was better than the first, though it lay by
him several years, excited some interest at Cambridge;
and Sir Isaac himself informs us, that one of the Fellows
of Trinity College had completed a telescope of the same
kind, which he considered as somewhat superior to his
own. The existence of these telescopes having become
known to the Royal Society, Newton was requested to
send his instrument for examination to that learned body.
He accordingly transmitted it to Mr. Oldenburg in December, 1671, and from this epoch his name began to
acquire that celebrity by which it has been so peculiarly
distinguished.

On the 11th of January, 1671, it was announced to the
Royal Society, that his reflecting telescope had been shown
to the King, and had been examined by the president, Sir
Robert Moray, Sir Paul Neale, Sir Christopher Wren,
and Mr. Hook. These gentlemen entertained so high an
opinion of it, that, in order to secure the honour of the
contrivance to its author, they advised the inventor to
send a drawing and description of it to Mr. Huygens at
Paris. Mr. Oldenburg accordingly drew up a description
of it in Latin, which, after being corrected by Mr. Newton,

was transmitted to that eminent philosopher. This telescope, of which the annexed is an accurate drawing, is carefully preserved in the library of the Royal Society of London, with the following inscription :—

"*Invented by Sir Isaac Newton, and made with his own hands*, 1671."

It does not appear that Newton executed any other reflecting telescopes than the two we have mentioned. He informs us that he re-polished and greatly improved a fourteen feet object-glass executed by a London artist; and having proposed in 1678 to substitute glass reflectors in place of metallic specula, he tried to make a reflecting telescope on this principle four feet long, and with a magnifying power of 150. The glass was wrought by a London artist; and, though it seemed well finished, yet, when it was quicksilvered on its convex side, it exhibited all over the glass innumerable inequalities, which gave an indistinctness to every object. He expresses, however, his conviction that nothing but good workmanship is wanting to perfect these telescopes, and he recommends their consideration "to the curious in figuring glasses."

For a period of fifty years this recommendation excited no notice. At last Mr. James Short of Edinburgh, an artist of consummate skill, executed, about the year 1730, no fewer than six reflecting telescopes with glass specula, three of fifteen inches, and three of nine inches in focal length. He found it extremely troublesome to give them a true figure with parallel surfaces; and several of them when finished turned out useless, in consequence of the veins which then appeared in the glass. Although these instruments performed remarkably well, yet the light was fainter than he expected, and from this cause, combined with the difficulty of finishing them, he afterwards devoted his labours solely to those with metallic specula.

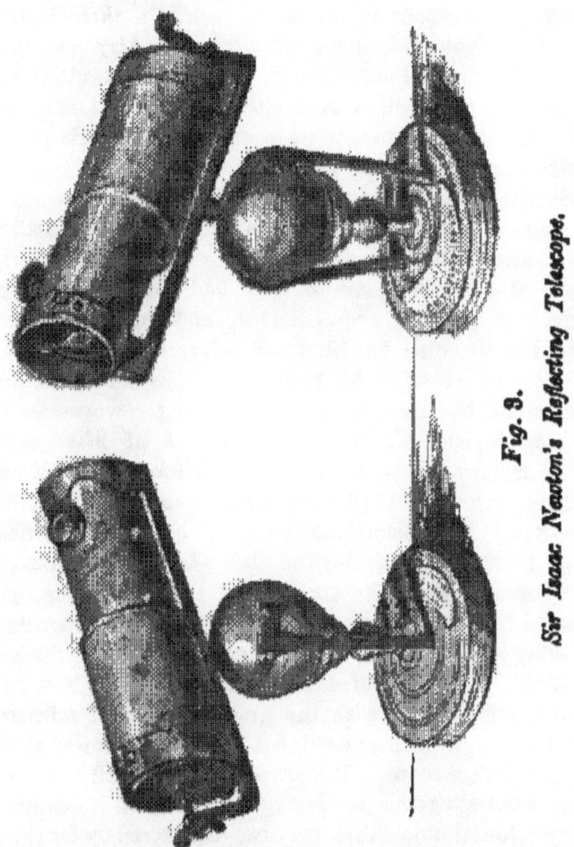

At a later period, in 1822, Mr. G. B. Airy* of Trinity College, and one of the distinguished successors of Newton in the Lucasian Chair, resumed the consideration of glass specula, and demonstrated that the aberration both of

* Now Sir George Airy, K C.B., and, since 1835, Astronomer Royal. —Editor.

figure and of colour might be corrected in these instruments. Upon this ingenious principle Airy executed more than one telescope; but, though the result of the experiment was such as to excite hopes of ultimate success, yet the construction of such instruments is still a desideratum in practical science.*

Such were the attempts which Sir Isaac Newton made to construct reflecting telescopes; but nothwithstanding the success of his labours, neither the philosopher nor the practical optician seems to have had courage to pursue them. A London artist, indeed, undertook to imitate these instruments; but Sir Isaac informs us that "he fell much short of what he had attained, as he afterwards understood by discoursing with the under workman he had employed." After a long period of fifty years, John Hadley, Esq., of Essex, a Fellow of the Royal Society, began in 1719 or 1720 to execute a reflecting telescope. His scientific knowledge and his manual dexterity fitted him admirably for such a task; and, probably after many failures, he constructed two large telescopes about 5 feet 3 inches long, one of which, with a speculum 6 inches in diameter, was presented to the Royal Society in 1723. The celebrated Dr. Bradley and the Rev. Mr. Pound compared it with the great Huygenian refractor 123 feet long. It bore as high a magnifying power as the Huygenian telescope. It showed objects equally distinct, though not altogether so clear and bright, and it exhibited every celestial object that had been discovered by Huygens —the five satellites of Saturn, the shadow of Jupiter's satellites on his disc, the black list in Saturn's ring, and

* Since this was written, Mr. With of Hereford has constructed a large number of silvered glass specula for telescopes, and we have had several opportunities of witnessing the excellence of their performance. They are mounted, and sold by Mr. Browning of the Minories, London, whose constructive ability has conferred so many benefits upon astronomical science. —EDITOR.

the edge of his shadow cast on the ring. Encouraged and instructed by Mr. Hadley, Dr. Bradley began the construction of reflecting telescopes, and succeeded so well that he would have completed one of them, had he not been obliged to change his residence. Some time afterwards he and the Hon. Samuel Molyneux undertook the task together at Kew, and attempted to execute specula about twenty-six inches in focal length; but notwithstanding Dr. Bradley's former experience, and Mr. Hadley's frequent instructions, it was a long time before they succeeded. The first good instrument which they finished was in May, 1724. It was twenty-six inches in focal length; but they afterwards completed a very large one of eight feet, the largest that had ever been made. The first of these instruments was afterwards elegantly fitted up by Mr. Molyneux, and presented to his Majesty John V., King of Portugal.

The great object of these two able astronomers was to reduce the method of making specula to such a degree of certainty that they could be manufactured for public sale. Mr. Hawksbee had indeed made a good one about three and a half feet long, and had proceeded to the execution of two others, one of six feet, and another of twelve feet in focal length; but Mr. Scarlet and Mr. Hearne, having received all the information which Mr. Molyneux had acquired, constructed them for public sale; and the reflecting telescope has ever since been an article of trade with every regular optician.

As Sir Isaac Newton was at this time President of the Royal Society, he had the high satisfaction of seeing his own invention become an instrument of public use, and of great advantage to science, and he no doubt felt the full influence of this triumph of his skill. Still, however, the reflecting telescope had not achieved any new discovery in the heavens. The latest accession to astronomy had been

made by the ordinary refractors of Huygens, labouring
under all the imperfections of coloured light; and this
long pause in astronomical discovery seemed to indicate
that man had carried to its farthest limits his power of
penetrating into the depths of the universe. This, however,
was only one of those stationary positions from which
human genius takes a new and a loftier elevation. While
the English opticians were thus practising the recent art of
grinding specula, Mr. James Short, of Edinburgh, was
devoting to the subject all the energies of his youthful
mind. In 1732, and in the twenty-second year of his age,
he began his labours, and he carried to such high perfection the art of grinding and polishing specula, and of
giving them the true parabolic figure, that, with a telescope
fifteen inches in focal length, he read in the Philosophical
Transactions at the distance of 500 feet, and frequently
saw the five satellites of Saturn together, a power which
was beyond the reach even of Hadley's six foot instrument.
The celebrated Maclaurin compared the telescopes of Short
with those made by the best London artists, and so great
was their superiority, that his small telescopes were invariably superior to larger ones from London. In 1742, after
he had settled as an optician in the metropolis, he executed
for Lord Thomas Spencer a reflecting telescope, twelve
foot in focal length, for £630; in 1752 he completed one
for the King of Spain, at the expense of £1,200; and a
short time before his death, which took place in 1768, he
finished the specula of the large telescope which was
mounted equatorially for the observatory of Edinburgh by
his brother Thomas Short, who was offered twelve hundred
guineas for it by the King of Denmark.

Although the superiority of these instruments, which
were all of the Gregorian form, demonstrated the value
of the reflecting telescope, yet no skilful hand had yet
directed it to the heavens; and it was reserved for a Her-

schel to be the first to employ it as an instrument of discovery, to exhibit to the eye of man new worlds and new systems, and to bring within the grasp of his reason those remote regions of space to which his imagination even had scarcely ventured to extend its power. So early as 1774 he completed a *five* foot Newtonian reflector, and he afterwards executed no fewer than *two hundred* 7 feet, *one hundred and fifty* 10 feet, and *eighty* 20 feet specula. In 1781 he began a reflector thirty feet long, and having a speculum thirty-six inches in diameter; and under the munificent patronage of George III. he completed, in 1789, his gigantic instrument forty foot long, with a speculum *forty-seven and a half* inches in diameter. The genius and perseverance which created instruments of such transcendent magnitude were not likely to terminate with their construction. In the examination of the starry heavens, the ultimate object of his labours, Sir W. Herschel exhibited the same exalted qualifications, and in a few years he rose from the level of humble life to the enjoyment of a name more glorious than that of the sages and warriors of ancient times, and as immortal as the objects with which it will be for ever associated. Nor was it in the ardour of the spring of life that those triumphs of reason were achieved. Sir William Herschel had reached the middle of his course before his career of discovery began, and it was in the autumn and winter of his days that he reaped the full harvest of his glory. The discovery of a new planet at the verge of the solar system was the first trophy of his skill; and new double and multiple stars, and new nebulæ and groups of celestial bodies, were added in thousands to the system of the universe. The spring-tide of knowledge which was thus let in upon the human mind, continued for a while to spread its waves over Europe; but when it sank to its ebb in England, there was no other bark left upon the strand but that of the Deucalion of Science, whose home had been so long upon its waters.

During the life of Herschel, and during the reign and within the dominions of his royal patron, four new planets were added to the solar system; but they were detected by telescopes of ordinary power; and we venture to state that, since the reign of George III., no attempt has been made to keep up the continuity of Sir William Herschel's discoveries.

What avails the enthusiasm and the efforts of individual minds in the intellectual rivalry of nations? When the proud science of England pines in obscurity, blighted by the absence of the royal favour, and of the nation's sympathy,—when its chivalry fall unwept and unhonoured,—how can it sustain the conflict against the honoured and marshalled genius of foreign lands ? *

* Brewster here seems to take too gloomy a view of the state of English science, and (we may add) too high a one of the advantages of "royal favour." Of late years, at any rate, the progress of scientific discovery, aided chiefly by private enterprise and amateur zeal, has been immense.— EDITOR.

CHAPTER IV.

He delivers a Course of Optical Lectures at Cambridge—Is elected Fellow of the Royal Society—He communicates to them his Discoveries on the Different Refrangibility and Nature of Light—Popular Account of them—They involve him in various Controversies—His Dispute with Purdies —Linus—Lucas—Dr. Hooke and Mr. Huygens—The Influence of these Disputes on the Mind of Newton.

ALTHOUGH Newton delivered a course of lectures on optics in the University of Cambridge in the years 1669, 1670, and 1671, containing his principal discoveries relative to the different refrangibility of light, yet it is a singular circumstance, that these discoveries should not have become public through the conversation or correspondence of his pupils. The Royal Society had acquired no knowledge of them until the beginning of 1672, and his reputation in that body was founded chiefly on his reflecting telescope. On the 23rd of December, 1671, the celebrated Dr. Seth Ward, Bishop of Salisbury, who was the author of several able works on astronomy, and had filled the Savilian chair of astronomy at Oxford, proposed Mr. Newton as a Fellow of the Royal Society. The satisfaction which he derived from this circumstance appears to have been considerable; and, in a letter to Mr. Oldenburg, of the 6th of January, he says, "I am very sensible of the honour done me by the Bishop of Sarum in proposing me a candidate; and which, I hope, will be further conferred upon me by my election into the Society; and if so, I shall endeavour to testify my gratitude, by communicating what my poor and solitary endeavours can effect towards the promoting your philosophical designs." His election accordingly took place on the 11th of January, the same day on which the Society agreed to transmit a description of his telescope to Mr. Huygens at Paris. The notice of his election, and the thanks of the Society for the communication of his telescope, were conveyed in the same letter, with an assurance that the Society "would take care that all right should be done in the matter of this invention." In his next letter to Oldenburg, written on the 18th of January, 1671-2, he announces his optical discoveries in the following remarkable manner: "I desire that in your next letter you would inform me for what time the Society continue their weekly meetings; because if they continue them for any time, I am purposing them, to be considered of and examined, an account of a philosophical discovery which induced me to the making of the said telescope; and I doubt not but will prove much more grateful than the communication of that instrument; being in my judgment the oddest, if not the most considerable, detection which hath hitherto been made in the operations of nature."

This "considerable detection" was the discovery of the different refrangibility of the rays of light which we have already explained, and which led to the construction of his reflecting telescope. It was communicated to the Royal Society in a letter to Mr. Oldenburg, dated February 6th, and excited great interest amongst its members. The "solemn thanks" of the meeting were ordered to be transmitted to its author for his "very ingenious discourse." A desire was expressed to have it immediately printed, both for the purpose of having it well considered by philosophers, and for "securing the considerable notices thereof to the author against the arrogations of others," and Dr. Seth Ward, Bishop of Salisbury, Mr. Boyle, and Dr. Hooke, were desired to peruse and consider it, and to bring in a report upon it to the Society.

The kindness of this distinguished body, and the anxiety which they had already evinced for his reputation, excited on the part of Newton a corresponding feeling, and he gladly accepted of their proposal to publish his discourse in the monthly numbers in which the Transactions were then given to the world. "It was an esteem," says he,* " of the Royal Society for most candid and able judges in philosophical matters, encouraged me to present them with that discourse of light and colours, which since they have so favourably accepted of, I do earnestly desire you to return them my cordial thanks. I before thought it a great favour to be made a member of that honourable body, but I am now more sensible of the advantages ; for believe me, Sir, I do not only esteem it a duty to concur with you in the promotion of real knowledge ; but a great privilege, that, instead of exposing discourses to a prejudiced and common multitude, (by which means many truths have been baffled and lost,) I may with freedom apply myself to so judicious and impartial an assembly. As to the printing of that letter, I am satisfied in their judgment, or else I should have thought it too strait and narrow for public view. I designed it only to those that know how to improve upon hints of things ; and therefore, to spare tediousness, omitted many such remarques and experiments as might be collected by considering the assigned laws of refractions ; some of which I believe, with the generality of men, would yet be almost as taking as any I described. But yet, since the Royal Society have thought it fit to appear publicly, I leave it to their pleasure : and perhaps to supply the aforesaid defects, I may

* Letter to Oldenburg, February 10th, 1671.

send you some more of the experiments to second it (if it be so thought fit) in the ensuing Transactions."

Following the order which Newton himself adopted, we have, in the preceding chapter, given an account of the leading doctrine of the different refrangibility of light, and of the attempts to improve the reflecting telescope which that discovery suggested. We shall now, therefore, endeavour to make the reader acquainted with the other discoveries respecting colours, which he at this time communicated to the Royal Society.

Having determined, by experiments already described, that a beam of white light, as emitted from the sun, consisted of seven different colours, which possess different degrees of refrangibility, he measured the relative extent of the coloured spaces, and found them to have the proportions shown in Fig. 4, which represents the *prismatic spectrum*, and which is nothing more than an elongated image of the sun, produced by the rays being separated in different degrees from their original direction, the *red* being refracted *least*, and the *violet most* powerfully.

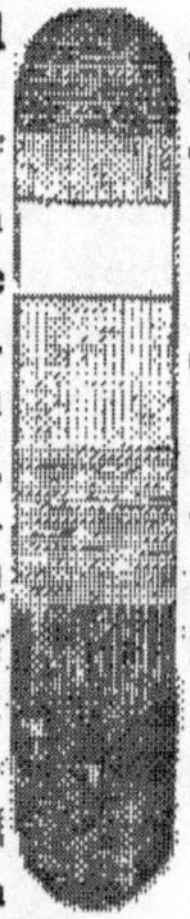

If we consider light as consisting of minute particles of matter, we may form some notion of its decomposition by the prism from the following popular illustration. If we take steel filings of seven different degrees of fineness, and mix them together, there are two ways in which we may conceive the mass to be decomposed; or, what is the same thing, all the seven different kinds of filings separated from each other. By means of seven sieves of different degrees of fineness, and so made that the finest will just transmit the finest powder, and detain all the rest, while the next in fineness transmits the two finest powders and detains all the rest, and so on, it is obvious that all the powders may be completely separated from each other. If we again mix all the steel filings, and, laying them upon a table, hold high above them a flat bar magnet, so that none of the filings are attracted, then if we bring the magnet nearer and nearer, we shall come to a point where the finest filings are drawn up to it. These being removed, and the magnet brought nearer still, the next finest powders will be attracted, and so on till we have thus drawn out of the mass all the powders in a separate state. We may conceive the bar magnet to be inclined to the surface of the steel filings, and so moved over the mass, that at the end nearest to them the heaviest or coarsest will be attracted, and at the remotest end the finest or lighter filings, while the rest are attracted to intermediate points, so that the seven different filings are not only separated, but are found adhering in separate patches to the surface of the flat magnet. The first of these methods with the sieves may represent the process of decomposing light by which certain rays of white light are absorbed, or stifled, or stopped in passing through bodies while certain other rays are transmitted. The second method may represent the process of decomposing light by refraction, or by the attraction of certain rays farther from their original direction than other rays; and the different patches of filings upon the flat magnet may represent the spaces on the spectrum.

When a beam of white light is decomposed into the seven different colours of the spectrum, any particular colour, when once separated from the rest, is not susceptible of any change, or farther decomposition, whether it is refracted through prisms or reflected from mirrors. It may become fainter or brighter, but Newton never could, by any process, alter its colour or its refrangibility.

Among the various bodies which act upon light, it is

conceivable that there might have been some which acted least upon the violet rays and most upon the red rays. Newton, however, found that this never took place ; but that the same degree of refrangibility always belonged to the same colour, and the same colour to the same degree of refrangibility.

Having thus determined that the seven different colours of the spectrum were original or simple, he was led to the conclusion that *whiteness* or white light is a compound of all the seven colours of the spectrum, in the proportions in which they are represented in Fig. 4.   In order to prove this, or what is called the recomposition of white light out of the seven colours, he employed three different methods.

When the beam RR (Fig. 1) was separated into its

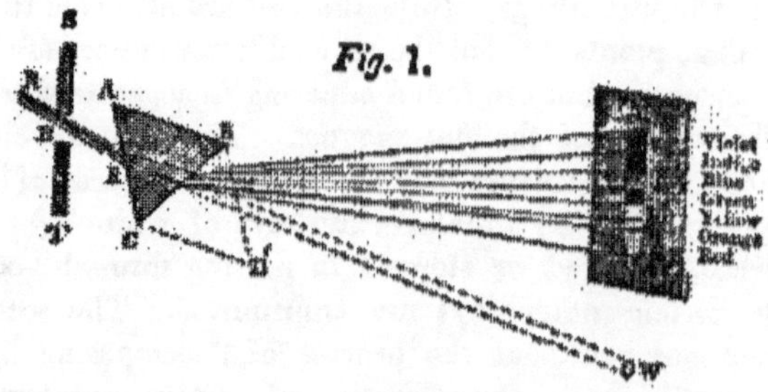

Fig. 1.

elementary colours by the prism ABC, he received the colours on another prism BCD', held either close to the first or a little behind it, and by the opposite refraction of this prism they were all refracted back into a beam of white light OW, which formed a white circular image on the wall at W, similar to what took place before any of the prisms were placed in its way.

The other method of recomposing white light consisted in making the spectrum fall upon a lens at some distance from it.   When a sheet of white paper was held behind the lens, and removed to a proper distance, the colours were all refracted into a circular spot, and so blended as to reproduce light so perfectly white as not to differ sensibly from the direct light of the sun.

The last method of recomposing white light was one more suited to vulgar apprehension. It consisted in attempting to compound a white by mixing the coloured powders used by painters. He was aware that such colours, from their very nature, could not compose a pure white; but even this imperfection in the experiment he removed by an ingenious device. He accordingly mixed one part of *red lead*, four parts of *blue bise*, and a proper proportion of *orpiment* and *verdigris*. This mixture was *dun*, like wool newly cut, or like the human skin. He now took one-third of the mixture and rubbed it thickly on the floor of his room, where the sun shone upon it through the opened casement, and beside it, in the shadow, he laid a piece of white paper of the same size. " Then going from them to the distance of twelve or eighteen feet, so that he could not discern the unevenness of the surface of the powder, nor the little shadows let fall from the gritty particles thereof; the powder appeared intensely white, so as to transcend even the paper itself in whiteness." By adjusting the relative illumination of the powders and the paper, he was able to make them both appear of the very same degree of whiteness. " For," says he, " when I was trying this, a friend coming to visit me, I stopped him at the door, and before I told him what the colours were, or what I was doing, I asked him which of the two whites was the best, and wherein they differed? And after he had at that distance viewed them well, he answered, that they were both good whites, and that he could not say which was best, nor wherein their colours differed." Hence Newton inferred that perfect whiteness may be compounded of different colours.

As all the various shades of colour which appear in the

material world can be imitated by intercepting certain rays in the spectrum, and uniting all the rest, and as bodies always appear of the same colour as the light in which they are placed, he concluded that the colours of natural bodies are not qualities inherent in the bodies themselves, but arise from the disposition of the particles of each body to stop or absorb certain rays, and thus to reflect more copiously the rays which are not thus absorbed.

No sooner were these discoveries given to the world than they were opposed with a degree of virulence and ignorance which have seldom been combined in scientific controversy.　　Unfortunately for Newton, the Royal Society contained few individuals of pre-eminent talent, capable of appreciating the truth of his discoveries, and of protecting him against the shafts of his envious and ignorant assailants.　This eminent body, while they held his labours in the highest esteem, were still of opinion that his discoveries were fair subjects of discussion; and their Secretary accordingly communicated to him all the papers which were written in opposition to his views.　The first communication on the subject was the suggestion of four experiments with the prism, which seems to have been made by some friend at Cambridge, as Newton communicated them to the Editor of the Philosophical Transactions with his own observations (on April 13, 1672).　This letter was followed by a communication from a Jesuit named Ignatius Pardies, Professor of Mathematics at Clermont, who pretended that the elongation of the sun's image arose from the unequal incidence of the different rays on the first face of the prism, although Newton had demonstrated in his own discourse that this was not the case. In April, 1672, Newton transmitted to Oldenburg a decisive reply to the animadversions of Pardies; but, unwilling to be vanquished, this disciple of Descartes took up a fresh position, and maintained that the elongation of the spectrum might be explained by the diffusion of light on the hypothesis of Grimaldi, or by the diffusion of undulations on the hypothesis of Hooke. Newton again replied to these feeble reasonings; but he contented himself with reiterating his original experiments, and confirming them by more popular arguments, with which Pardies at last professed himself to be satisfied, and wrote on the 9th of July in terms highly complimentary to Newton.

Another combatant soon sprang up in the person of one Francis Linus, a physician in Liège,* who, on the 6th of October, 1674, addressed a letter to a friend in London, containing animadversions on Newton's doctrine of colours. He boldly affirms that, in a perfectly clear sky, the image of the sun made by a prism is never elongated, and that the spectrum observed by Newton was not formed by the true sunbeams, but by rays proceeding from some bright cloud. In support of these assertions, he appeals to frequently repeated experiments on the refractions and reflections of light which he had exhibited thirty years before to Sir Kenelm Digby, " who took notes upon them ;" and he unblushingly states that, if Newton had used the same industry as he did, he would never have " taken so impossible a task in hand, as to explain the difference between the length and breadth of the spectrum by the received laws of refraction." When this letter was shown to Newton, he refused to answer it; but a letter was sent to Linus, referring him to the answer to Pardies, and assuring him that the experiments on the spectrum were made when there was no bright cloud in the heavens. This

* This gentleman was the author of a paper in the Philosophical Transactions, entitled, " Optical Assertions concerning the Rainbow." How such a paper could be published by so learned a body seems in the present day utterly incomprehensible. The dials which Linus erected at Liège, and which were the originals of those formerly in the Priory Gardens in London, are noticed in the Philosophical Transactions for 1703. In one of them the hours were distinguished by touch.

reply, however, did not satisfy the Dutch experimentalist. On the 25th of February, 1675, he addressed another letter to his friend, in which he gravely attempts to prove that the experiment of Newton was not made in a clear day; that the prism was not close to the hole; and that the length of the spectrum was not perpendicular, or parallel to the length of the prism. Such assertions could not but irritate even the patient mind of Newton. He more than once declined the earnest request of Oldenburg to answer these observations; he stated, that, as the dispute referred to matters of fact, it could only be decided before competent witnesses, and he referred to the testimony of those who had seen his experiments. The entreaties of Oldenburg, however, prevailed over his own better judgment; and, "lest Mr. Linus should make the more stir," this great man was compelled to draw up a long and explanatory reply to reasonings utterly contemptible, and to assertions altogether unfounded. This answer, dated November 13th, 1675, could scarcely have been perused by Linus, who was dead on the 15th of December, when his pupil, Mr. Gascoigne, took up the gauntlet, and declared that Linus had shown to various persons in Liège the experiment, which proved the spectrum to be circular, and that Newton could not be more confident on his side than they were on the other. He admitted, however, that the different results might arise from different ways of placing the prism. Pleased with the "handsome genius of Mr. Gascoigne's letter," Newton replied even to it, and suggested that the spectrum seen by Linus may have been the circular one, formed by one reflexion; or, what he thought more probable, the circular one, formed by two refractions and one intervening reflexion from the base of the prism, which would be coloured if the prism was not an isosceles one. This suggestion seems to have enlightened the Dutch philosophers. Mr. Gascoigne, having no conveniences for making the experiments pointed out by Newton, requested

Mr. Lucas of Liège to perform them in his own house. This ingenious individual, whose paper gave great satisfaction to Newton, and deserves the highest praise, confirmed the leading results of the English philosopher; but though the refracting angle of his prism was 60°, and the refractive power 1·500, he never could obtain a spectrum whose length was more than from *three* to *three-and-a-half* times its breadth, when the refractions on both sides of the prism were equal, while Newton found the length to be *five* times its breadth. In our author's reply, he directs his attention principally to this point of difference. He repeated his measures with each of the three angles of *three* different prisms, and he affirmed that Mr. Lucas might *make sure to find the image as long or longer than he had yet done*, by taking a prism with plain surfaces, and with an angle of 66° or 67°. He admitted that the smallness of the angle in Mr. Lucas's prism, viz., 60°, did not account for the shortness of the spectrum which he obtained with it; and he observed in one of his own prisms that the length of the image was greater in proportion to the refracting angle than it should have been: an effect which he ascribes to its having a greater refractive power. There is every reason to believe that the prism of Lucas had actually a less dispersive power than that of Newton; and had the Dutch philosopher measured its refractive power, instead of guessing it, or had Newton been less confident than he was,* that all other prisms must give a spectrum

* Newton speaks with singular positiveness on this subject. "For *I know*," says he, "that Mr. Lucas's observations *cannot hold* where the refracting angle of the prism is full 60°, and the day is clear, and the full length of the colours is measured, and the breadth of the image answers to the sun's diameter: and seeing I am well assured of the truth and exactness of my own observations, I shall be unwilling to be diverted by any other experiments from having a fair end made of this in the first place." On the supposition that his prism was one of very low dispersive power, Mr. Lucas might, with perfect truth, have used the very same language towards Newton,

of the same length as his in relation to its refracting angle and its index of refraction, the invention of the achromatic telescope would have been the necessary result. The objections of Lucas drove our author to experiments which he had never before made,—to measure accurately the lengths of the spectra with different prisms of different angles and different refractive powers; and had the Dutch philosopher maintained his position with more obstinacy, he would have conferred a distinguished favour upon science, and would have rewarded Newton for all the vexation which had sprung from the minute discussion of his optical experiments.

Such was the termination of his disputes with the Dutch philosophers, and it can scarcely be doubted that it cost him more trouble to detect the origin of his adversaries' blunders, than to establish the great truths which they had attempted to overturn.

Harassing as such a controversy must have been to a philosopher like Newton, yet it did not touch those deep-seated feelings which characterize the noble and generous mind. No rival jealousy yet pointed the arguments of his opponents;—no charges of plagiarism were yet directed against his personal character. These aggravations of scientific controversy, however, he was destined to endure; and in the dispute which he was called to maintain both against Hooke and Huygens, the agreeable consciousness of grappling with men of kindred powers was painfully embittered by the personality and jealousy with which it was conducted.

Dr. Robert Hooke was about seven years older than Newton, and was one of the ninety-eight original or unelected members of the Royal Society. He possessed great versatility of talent; yet, though his genius was of the most original cast, and his acquirements extensive, he had not devoted himself with fixed purpose to any particular branch of knowledge. His numerous and ingenious inventions, of which it is impossible to speak too highly, gave to his studies a practical turn which unfitted him for that continuous labour which physical researches so imperiously demand. The subjects of light, however, and of gravitation, seem to have deeply occupied his thoughts before Newton appeared in the same field, and there can be no doubt that he had made considerable progress in both of these inquiries. With a mind less divergent in its pursuits, and more endowed with patience of thought, he might have unveiled the mysteries in which both these subjects were enveloped, and pre-occupied the intellectual throne which was destined for his rival; but the infirm state of his health—the peevishness of temper which this occasioned—the number of unfinished inventions from which he looked both for fortune and fame—and, above all, his inordinate love of reputation—distracted and broke down the energies of his powerful intellect. In the more matured inquiries of his rivals, he recognised, and often truly, his own incompleted speculations; and when he saw others reaping the harvest for which he had prepared the ground, and of which he had sown the seeds, it was not easy to suppress the mortification which their success inspired. In the history of science, it has always been a difficult task to adjust the rival claims of competitors, when the one was allowed to have completed what the other was acknowledged to have begun. He who commences an inquiry, and publishes his results, often goes much farther than he has announced to the world, and, pushing his speculations into the very heart of the subject, frequently submits them to the ear of friendship. From the pedestal of his published labours his rival begins his researches, and brings them to a successful issue; while he has in reality done nothing more than complete and demonstrate the imperfect speculations of his predecessor. To the world, and to himself, he is no doubt in the position of the principal discoverer; but there is still some apology for his rival when he brings forward his unpublished labours; and some excuse for the exercise of personal feeling, when he measures the speed of his rival by his own proximity to the goal.

The conduct of Dr. Hooke would have been viewed with some such feeling, had not his arrogance on other occasions checked the natural current of our sympathy. When Newton presented his reflecting telescope to the Royal Society, Dr. Hooke not only criticized the instrument with undue severity, but announced that he possessed an infallible method of perfecting all kinds of optical instruments, so that "whatever almost hath been in notion and imagination, or desired in optics, may be performed with great facility and truth."

Hooke had been strongly impressed with the belief (derived, with some modifications, from the theory of Descartes,) that light consisted in the undulations of a highly elastic medium pervading all bodies; and guided by his experimental investigation of the phenomena of diffraction, he had even announced the great *principle of interference*, which has performed such an important part in modern science. Regarding himself, therefore, as in possession of the true theory of light, he examined the discoveries of Newton in their relation to his own speculative views; and, finding that their author was disposed to consider that element as consisting of material particles, he did not scruple to reject doctrines which he believed to be incompatible with truth. Dr. Hooke was too accurate an observer, not to admit the general correctness of Newton's observations. He allowed the existence of different refractions, the unchangeableness of the simple colours, and the production of white light by the union of all the colours of the spectrum; but he maintained that the

different refractions arose from the splitting and rarefying
of ethereal pulses, and that there are only two colours in
nature, viz., *red* and *violet*, which produce by their mixture
all the rest, and which are themselves formed by the two
sides of a split pulse or undulation.

In reply to these observations, Newton wrote an able
letter to Oldenburg, dated June 11th, 1672, in which he
examined with great boldness and force of argument the
various objections of his opponent, and maintained the
truth of his doctrine of colours, as independent of the two
hypotheses respecting the origin and production of light.
He acknowledged his own partiality to the doctrine of the
materiality of light; he pointed out the defects of the
undulatory theory; he brought forward new experiments
in confirmation of his former results; and he refuted the
opinions of Hooke respecting the existence of only two
simple colours.   No reply was made to the powerful arguments of Newton; and Hooke contented himself with laying
before the Society his curious observations on the colours
of soap bubbles, and of plates of air, and in pursuing his
experiments on the diffraction of light, which, after an
interval of two years, he laid before the same body.

After he had thus silenced the most powerful of his
adversaries, Newton was again called upon to defend himself against a new enemy. Christian Huygens, an
eminent mathematician and natural philosopher, who, like
Hooke, had maintained the undulatory theory of light,
transmitted to Oldenburg various animadversions on the
Newtonian doctrine; but though his knowledge of optics
was of the most extensive kind, yet his objections were
nearly as groundless as those of his less enlightened
countrymen. Attached to his own hypothesis respecting
the nature of light, namely, to the system of undulation,
he seems, like Dr. Hooke, to have regarded the discoveries
of Newton as calculated to overturn it; but his principal

objections related to the composition of colours, and
particularly of white light, which he alleged could be
obtained from the union of two colours, *yellow* and *blue*.
To this and similar objections, Newton replied that the
colours in question were not simple *yellow* and *blue*, but
were compound colours, in which, together, all the colours
of the spectrum were themselves blended; and though he
evinced some strong traces of feeling at being again put
upon his defence, yet his high respect for Huygens induced
him to enter with patience on a fresh development of his
doctrine. Huygens felt the reproof which the tone of
this answer so gently conveyed, and in writing to Oldenburg he used the expression, that Mr. Newton "maintained his doctrine with some concern." To this our
author replies, " As for Mr. Huygens's expression, I confess it was a little ungrateful to me, to meet with objections
which had been answered before, without having the least
reason given me why those answers were insufficient."
But though Huygens appears in this controversy as a rash
objector to the Newtonian doctrine, it was afterwards the
fate of Newton to play a similar part against the Dutch
philosopher. When Huygens published his beautiful law
of double refraction in Iceland spar, founded on the finest
experimental analysis of the phenomena, though presented
as a result of the undulatory system, Newton not only
rejected it, but substituted for it another law entirely
inconsistent not merely with the experiments of Huygens,
which Newton himself had praised, but with those of all
succeeding philosophers.

The influence of these controversies on the mind of
Newton seems to have been highly exciting. Even the
satisfaction of humbling all his antagonists he did not feel
as a sufficient compensation for the disturbance of his
tranquillity. "I intend," says he,* "to be no further

* Letter to Oldenburg in 1672, containing his first reply to Huygens.

solicitous about matters of philosophy.   And therefore I hope you will not take it ill if you find me never doing any thing more in that kind; or rather that you will favour me in my determination, by preventing, so far as you can conveniently, any objections or other philosophical letters that may concern me."   In a subsequent letter in 1675, he says, " I had some thoughts of writing a further discourse about colours, to be read at one of your assemblies; but find it yet against the grain to put pen to paper any more on that subject;" and in a letter to Leibnitz, dated December the 9th, 1675, he observes, " I was so persecuted with discussions arising from the publication of my theory of light, that I blamed my own imprudence for parting with so substantial a blessing as my quiet to run after a shadow."

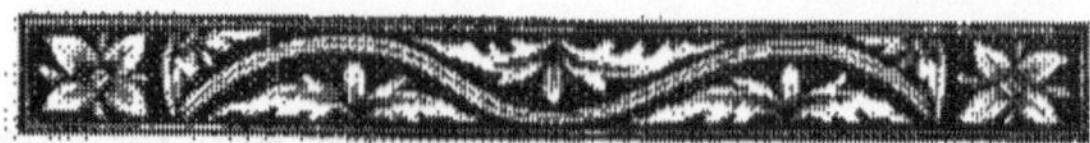

CHAPTER V.

THE new doctrines of the composition of light, and of the different refrangibility of the rays which compose it, having been thus established upon an impregnable basis, it will be interesting to take a general view of the changes which they have undergone since the time of Newton, and of their influence on the progress of optical discovery.

There are few facts in the history of science more singular than that Newton should have believed that all bodies, when shaped into prisms, produced prismatic spectra of equal length, or separated the red and violet rays to equal distances when the mean refraction, or the refraction of the middle ray of the spectrum, was the same. This opinion, which he deduced from no direct experiments and into which no theoretical views could have led him, seems to have been impressed upon his mind with all the force of an axiom.*   Even the shortness of the spec-

* In an experiment made by Newton, he had occasion to counteract the refraction of a prism of *glass* by another prism of *water;* and had he completed the experiment, and studied the result of it, he could not have failed to observe a quantity of uncorrected colour, which would have led him to the discovery of the different dispersive powers of bodies. But in order to increase the refractive power of the water, he mixed with it a little sugar of lead, the high dispersive power of which seems to have rendered the dispersive power of the water equal to that of the glass, and thus to have corrected the uncompensated colour of the glass prism.

trum observed by Lucas did not rouse him to farther inquiry; and when, under the influence of this blind conviction, he pronounced the improvement of the refracting telescope to be desperate, he checked for a long time the progress of this branch of science, and furnished to future philosophers a lesson which cannot be too deeply studied.

In 1729, about two years after the death of Sir Isaac, an individual unknown to science broke the spell in which the subject of the spectrum had been so singularly bound. Mr. Chester More Hall, of More Hall in Essex, while studying the mechanism of the human eye, was led to suppose that telescopes might be improved by a combination of lenses of different refractive powers, and he actually completed several object-glasses upon this principle. The steps by which he arrived at such a construction have not been recorded; but it is obvious that he must have discovered what escaped the sagacity of Newton, that prisms made of different kinds of glass produced different degrees of separation of the *red* and *violet* rays, or gave spectra of different lengths when the refraction of the middle ray of the spectrum was the same.

In order to explain how such a property led him to the construction of a *telescope without colour*, or an *achromatic telescope*, let us take a lens LL of *crown* or *plate glass*, whose focal length LY is about twelve inches. When

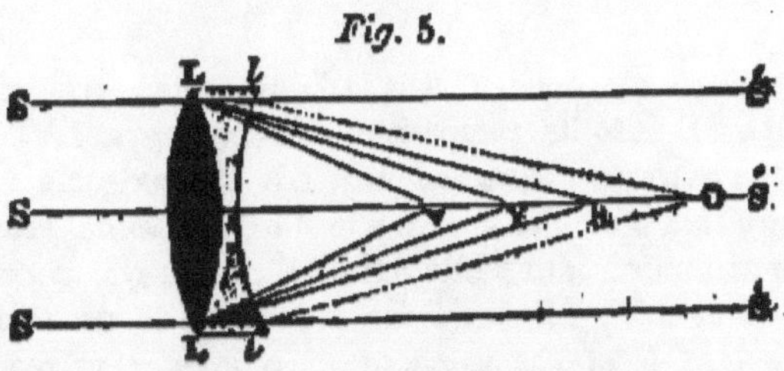

Fig. 5.

the sun's rays SL, SL fall upon it, the *red* will be refracted to R, the *yellow* to Y, and the *violet* to V. If we now place behind it a concave lens *ll* of the same glass, and of the same focus or curvature, it will be found, both by experiment and by drawing the refracted rays, according to the rules given in elementary works that the concave glass *ll* will refract the rays LR, LR into LS', LS', and the rays LV, LV into LS', LS', free of all colour ; but as these rays will be parallel, the two lenses will not have a focus, and consequently cannot form an image so as to be used as the object-glass of a telescope.   This is obvious from another consideration ; for since the curvatures of the convex and concave lenses are the same, the two put together will be exactly the same as if they were formed out of a single piece of glass, having parallel surfaces like a watch-glass, so that the parallel rays of light SL, SL will pass on in the same direction LS' LS' affected by equal and opposite refractions as in a piece of plane glass.

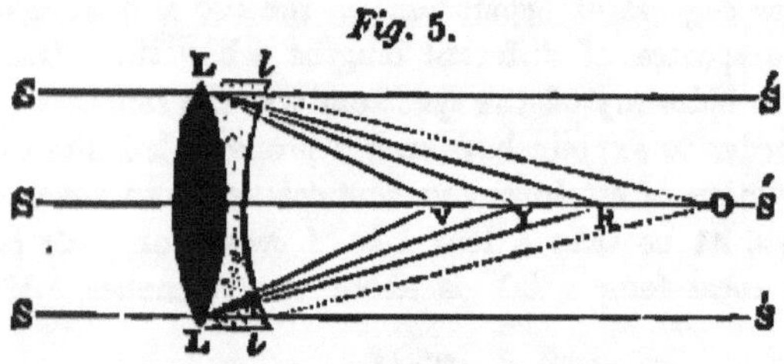

Fig. 5.

Now since the convex lens LL separated the white light SL, SL into its component coloured rays, LV, LV being the extreme violet, and LR, LR the extreme red ; it follows that a similar concave lens of the same glass is capable of uniting into white light LS', LS' rays, as much separated as LV, LR are. Consequently, if we take a concave lens *ll* of the same, or of a greater refractive power than the convex one, and having the power of uniting rays farther separated than LV, LR are, a less concavity in the lens ll will be sufficient to unite the rays LV, LR into a white ray LS'; but as the lens ll is now more concave than the lens LL is convex, the concavity will predominate, and the uncoloured rays LS', LS' will no longer be parallel, but will converge to some point O, where they will form a colourless or achromatic image of the sun.

The effect now described may be obtained by making the *convex* lens LL of *crown* or of *plate* glass, and the *concave* one of *flint* glass, or that of which wine glasses are made. If the concave lens ll has a greater refractive power than LL, which is always the case, the only effect of it will be to make the rays converge to a focus more remote than O, or to render a less curvature necessary in ll, if O is fixed for the focus of the combined lenses.

Such is the principle of the achromatic telescope, as constructed by Mr. Hall. This ingenious individual employed working opticians to grind his lenses, and he furnished them with the radii of the surfaces, which were adjusted to correct the aberration of figure as well as of colour. His invention, therefore, was not an accidental combination of a convex and a concave lens of different kinds of glass, which might have been made merely for experiment; but it was a complete achromatic telescope, founded on a thorough knowledge of the different dispersive powers of crown and flint glass. It is a curious circumstance, however, in the history of the telescope, that this invention was actually lost. Mr. Hall never published any account of his labours, and it is probable that he kept them secret till he should be able to present his instrument to the public in a more perfect form; and it was not till John Dollond had discovered the property of light upon which the instrument depends, and had actually constructed many fine telescopes, that the previous labours of Mr. Hall were laid before the public.* From this period the achromatic telescope underwent gradual improvement, and by the successive labours of Dollond, Ramsden, Blair, Tulley, Guinand, Lerebours, and Fraunhofer, it has become one of the most valuable instruments in physical science.

Although the achromatic telescope, as constructed by Dollond, was founded on the principle that the spectra formed by crown and flint glass differed only in their relative lengths, when the refraction of the mean ray was the same, yet by a more minute examination of the best instruments, it was found that they exhibited white or luminous objects tinged on one side with a green fringe, and on the other with one of a claret colour. These colours, which did not arise from any defect of skill in the artist, were found to arise from a difference in the extent of the coloured spaces in two equal spectra formed by crown and by flint glass. This property was called the *irrationality* of the coloured spaces ; and the uncorrected colours which remained when the primary spectrum of the crown glass was corrected by the primary spectrum of the flint glass, were called the *secondary* or *residual spectrum*. By a happy contrivance, which it would be out of place here to describe, Dr. Blair succeeded in correcting this secondary spectrum, or in removing the green and claret coloured fringes which appeared in the best telescopes, and to this contrivance he gave the name of the *Aplanatic Telescope.*

But while Newton thus overlooked these remarkable properties of the prismatic spectrum, as formed by different bodies, he committed some considerable mistakes in his examination of the spectrum, which was under his own

* See the article OPTICS in the *Edinburgh Encyclopædia*, vol. xv., p. 479, *note.*

immediate examination. It does not seem to have occurred
to him that the relations of the coloured spaces must be
greatly modified by the angular magnitude of the sun or
the luminous body or aperture from which the spectrum is
obtained; and misled by an apparent analogy between the
length of the coloured spaces and the divisions of a musical
chord,* he adopted the latter, as representing the proportion of the coloured spaces in every beam of white light.
Had two other observers, one situated in Mercury, and
the other in Jupiter, studied the prismatic spectrum of the
sun by the same instruments, and with the same sagacity
as Newton, it is demonstrable that they would have obtained very different results. On account of the apparent
magnitude of the sun in Mercury, the observer there would
obtain a spectrum entirely without *green*, having *red*,
orange, and *yellow* at one end, the *white* in the middle,
and terminated at the other end with *blue* and *violet*. The
observer in Jupiter would, on the contrary, have obtained
a spectrum in which the colours were much more condensed, and the pure colours more separated. On the
planets still further from the sun a spectrum exactly similar
would have been obtained, notwithstanding the greater
diminution of the sun's apparent diameter. It may now
be asked, Which of all these spectra are we to consider as
exhibiting the number and arrangement and extent of the
coloured spaces proper to be adopted as the true analysis
of a solar ray?

The spectrum observed by Newton has surely no claim
to our notice merely because it was observed upon the

* "This result was obtained," as Newton says, "by an assistant whose
eyes were more critical than mine, and who, by right lines drawn across
the spectrum, noted the confines of the colours. And this operation
being divers times repeated both on the same and on several papers, I
found that the observations agreed well enough with one another."—
OPTICS, part II., Prop. iii. p. 110.

surface of the earth. The spectrum obtained in Mercury affords no analysis at all of the incident beam, the colours being almost all compound, and not homogeneous, and that of Newton is liable to the same objection. Had Newton examined his spectrum under the very same circumstances in winter and in summer, he would have found the analysis of the beam more complete in summer, on account of the diminution of the sun's diameter; and, therefore, we are entitled to say that neither the number nor the extent of the coloured spaces, as given by Newton, are those which belong to the true prismatic spectrum.

The spectrum obtained in Jupiter and Saturn is the only one where the analysis is complete, as it is incapable of having its character altered by any farther diminution of the sun's diameter. Hence we are forced to conclude, not only that the number and extent of the primitive homogeneous colours, as given by Newton, are incorrect; but that if he had attempted to analyse some of the primitive tints in the spectrum, he would have found them decidedly composed of heterogeneous rays. There is one consequence of these observations which is somewhat interesting. A rainbow formed in summer, when the sun's diameter is least, must have its colours more condensed and homogeneous than in winter, when the size of its disc is a maximum; and when the upper or the lower limb of the sun is eclipsed, a rainbow formed at that time will lose entirely the yellow rays, and have the green and the red in perfect contact. For the same reason, a rainbow formed in Venus and Mercury will be destitute of green rays, and have a brilliant bow of white light separating two coloured arches, while in Mars, Jupiter, Saturn, and Uranus, the bow will exhibit only four homogeneous colours.

Had Newton received upon his prism a beam of light transmitted through a very narrow aperture, he would have anticipated Wollaston and Fraunhofer in their fine discovery of the lines in the prismatic spectrum.

From his analysis of the solar spectrum, Newton concluded, " that to the same degree of refrangibility ever belonged the same colour, and to the same colour ever belonged the same degree of refrangibility;" and hence he inferred, that *red, orange, yellow, green, blue, indigo,* and *violet,* were primary and simple colours. He admitted, indeed, that " the same colours in specie with these primary ones may be also produced by composition. For a mixture of *yellow* and *blue* makes *green,* and of *red* and *yellow* makes *orange;*" but such compound colours were easily distinguished from the simple colours of the spectrum by the circumstance, that they are always capable of being resolved by the action of the prism into the two colours which compose them.

This view of the composition of the spectrum might have long remained unchallenged, had we not been able to apply to it a new mode of analysis. Though we cannot separate the *green* rays of the spectrum into *yellow* and *blue* by the refraction of prisms, yet if we possessed any substance which had a specific attraction for *blue* rays, and which stopped them in their course, and allowed the *yellow* rays to pass, we should thus analyse the *green* as effectually as if they were separated by refraction. The substance which possesses this property is a purplish blue glass, similar to that of which finger-glasses are made. When we view through a piece of this glass, about the twentieth of an inch thick, a brilliant prismatic spectrum, we find that it has exercised a most extraordinary absorptive action on the different colours which compose it. The *red* part of the spectrum is divided into *two red* spaces, separated by an interval entirely devoid of light. Next to the inner red space comes a space of bright *yellow,* separated from the red by a visible interval. After the yellow comes the *green*, with an obscure space between them, then follows the *blue* and the *violet*, the last of which has suffered little or no diminution. Now it is very obvious, that in this experiment the blue glass has actually absorbed the *red* rays, which, when mixed with the *yellow* on one side, constituted *orange*, and the *blue* rays, which, when mixed with the *yellow* on the other side, constituted *green*, so that the insulation of the *yellow* rays thus effected, and the disappearance of the *orange*, and of the greater part of the *green* light, proves beyond a doubt that the *orange* and *green* colours in the spectrum are compound colours, the former consisting of *red* and *yellow* rays, and the latter of *yellow* and *blue* rays *of the very same refrangibility*. If we compare the two red spaces of the spectrum seen through the blue glass with the red space seen without the blue glass, it will be obvious that the red has experienced such an alteration in its tint by the action of the blue glass, as would be effected by the absorption of a small portion of yellow rays; and hence we conclude, that the red of the spectrum contains a slight tinge of yellow, and that the yellow space extends over more than one-half of the spectrum, including the *red*, *orange*, *yellow*, *green*, and *blue* spaces.

I have found also that red light exists in the yellow space, and it is certain that, in the violet space, red light exists in a state of combination with the blue rays. From these and other facts, which it would be out of place here to enumerate, I have been led to the conclusion, that the prismatic spectrum consists of three different spectra, viz., red, yellow, and blue, all having the same length, all superposed, and each having its maximum intensity at the point where it predominates in the combined spectrum. Hence red, yellow, and blue rays of the very same refrangibility co-exist at every point of the spectrum; but the colour at any one point will be that of the predominant ray, and will depend upon the relative distance of the point from the maximum ordinate of the curve which represents the intensity of the light of each of the three spectra.

This structure of the spectrum, which harmonizes with the old hypothesis of three simple colours, will be understood from the annexed diagram, where MN is the spectrum of seven colours, all compounded of the three simple ones, *red*, *yellow*, and *blue*. The ordinates of the curves R, Y, and B, will express the intensities of each colour at different points of the spectrum. At the red extremity M of the spectrum, the pure *red* is scarcely altered by the

Fig. 6.

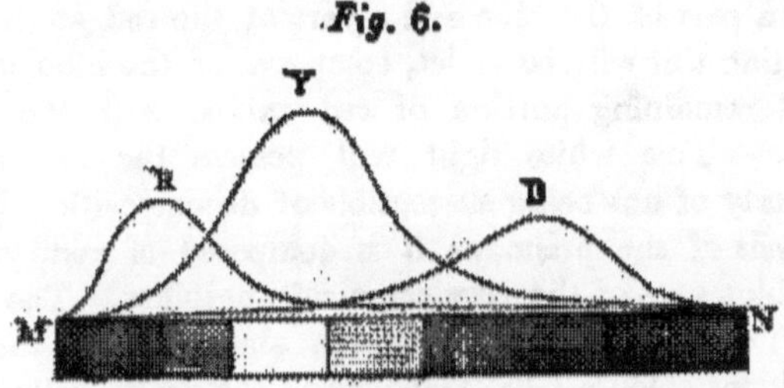

very slight intermixture of yellow and blue. Farther on in the red space, the *yellow* begins to make the red incline to scarlet. It then exists in sufficient quantity to form orange, and, as the red declines, the yellow predominates over the feeble portion of red and blue which are mixed with it. As the yellow decreases in intensity, the increasing blue forms with it a good green, and the blue rising to its maximum speedily overpowers the small portion of yellow and red. When the blue becomes very faint, the red exhibits its influence in converting it into violet, and the yellow ceases to exercise a marked influence on the tint. The influence of the red over the blue space is scarcely perceptible, on account of the great intensity of the blue light; but we may easily conceive it to reappear and form the violet light, not only from the rapid decline of the blue light, but from the greater influence of the red rays upon the retina.

These views may perhaps be more clearly understood by supposing that a certain portion of white light is actually formed at every point of the spectrum by the union of the requisite number of the three coloured rays that exist at any point. The white light thus formed will add to the brilliancy without affecting the tint of the predominant colour. In the violet space we may conceive the small portion of yellow which exists there to form white light with a part of the blue and a part of the red, so that the resulting tint will be violet, composed of the blue and the small remaining portion of red, mixed with the white light. This white light will possess the remarkable property of not being susceptible of decomposition by the analysis of the prism, as it is composed of red, yellow, and blue rays of the very same refrangibility. The insulation of this white light by the absorption of the predominant colours I have effected in the green, yellow, and red spaces; and by the use of new absorbing media we may yet hope to exhibit it in some of the other colours, particularly in the brightest part of the blue space, where an obvious approximation to it takes place.

Among the most important modern discoveries respecting the spectrum we must enumerate that of fixed dark and coloured lines (alluded to above), which we owe to the sagacity of Dr. Wollaston and M. Fraunhofer. Two or three of these lines were discovered by Dr. Wollaston, but nearly six hundred have been detected by means of the fine prisms and the magnificent apparatus of the Bavarian optician. These lines are parallel to one another, and perpendicular to the length of the spectrum. The largest occupy a space from 5″ to 10″ in breadth. Sometimes they occur in well-defined lines, and at other times in groups; and in all spectra formed from solar light, they preserve the same order and intensity, and the same relative position to the coloured spaces, whatever be the nature of the prism by which they are produced. Hence these lines mark fixed points, by which the relative dispersive powers of different media may be ascertained with a degree of accuracy hitherto unknown in this branch of science. In the light of the fixed stars, and in that of artificial flames, a different system of lines is produced; and this system remains unaltered, whatever be the nature of the prism by which the spectrum is formed.

The most important fixed lines in the spectrum formed by light emitted from the sun, whether it is reflected from the sky, the clouds, or the moon, may be easily seen by looking at a narrow slit in the window shutter of a dark room, through a hollow prism formed of plates of parallel glass, and filled with any fluid of a considerable dispersive power. The slit should not greatly exceed the twentieth of an inch, and the eye should look through the thinnest edge of the prism where there is the least thickness of fluid. These lines I have found to be the boundaries of spaces within which the rays have particular affinities for particular bodies.*

* It would be out of place to enter farther into this subject here, and I leave it as left by Brewster. The study of these lines has become in recent years a science of itself, and brought astronomy into closer union with chemistry than could have been thought possible, by the discoveries it has led to concerning the chemical constitution not only of the sun, but of many of the still more distant bodies of the universe.—EDITOR.

CHAPTER VI.

*Colours of thin Plates first studied by Boyle and Hooke—
Newton determines the Law of their Production—His
Theory of Fits of easy Reflexion and Transmission—
Colours of Thick Plates.*

In examining the nature and origin of colours as the
component parts of white light, the attention of Newton
was directed to the curious subject of the colours of thin
plates, and to its application to explain the colours of
natural bodies. His earliest researches on this subject
were communicated, in his Discourse on Light and Colours,
to the Royal Society, on the 9th of December, 1675,[*]
and were read at subsequent meetings of that body. This
discourse contained fuller details respecting the composition
and decomposition of light than he had given in his letter
to Oldenburg, and was concluded with nine propositions,
showing how the colours of thin transparent plates stand
related to those of all natural bodies.

The colours of thin plates seem to have been first observed by Mr. Boyle. Dr. Hooke afterwards studied them
with some care, and gave a correct account of the leading
phenomena, as exhibited in the coloured rings upon soap
bubbles, and between plates of glass pressed together. He

<hr>

[*] A little before this time Newton appears to have been occupied with
a subject very different from his more usual pursuits—the planting of
fruit trees for the manufacture of cider. Our author, in the enlarged
edition of this work, published a curious letter, dated September 2, 1676,
too long to insert here, written by Newton to Oldenburg on this subject,
which he had found amongst his papers.—EDITOR.

recognised that the colour depended upon some certain thickness of the transparent plate; but he acknowledges that he had attempted in vain to discover the relation between the thickness of the plate and the colour which it produced.

Dr. Hooke succeeded in splitting a mineral substance called mica, into films of such extreme thinness as to give brilliant colours. One plate, for example, gave a yellow colour, another a blue colour, and the two together a deep purple; but, as plates which produced those colours were always less than the twelve-thousandth part of an inch thick, it was quite impracticable, by any contrivance yet discovered, to measure their thickness, and determine the law according to which the colour varied with the thickness of the film. Newton surmounted this difficulty by laying a double convex lens, the radius of curvature of each side of which was fifty feet, upon the flat surface of a plano-convex object-glass, and in this way he obtained a plate of air or of space varying from the thinnest possible edge at the centre of the object-glass where it touched the plane surface, to a considerable thickness at the circumference of the lens. When light was allowed to fall upon the object-glass, every different thickness of the plate of air between the object-glass gave different colours; so that the point where the two object-glasses touched one another was the centre of a number of concentric coloured rings. Now, as the curvature of the object-glass was known, it was easy to calculate the thickness of the plate of air at which any particular colour appeared, and thus to determine the law of the phenomena.

In order to understand how he proceeded, let CED be the convex surface of the one object-glass, and AEB the flat surface of the other. Let them touch at the point E, and let homogeneous *red* rays fall upon them, as shown in the figure. At the point of contact E, where the plate of

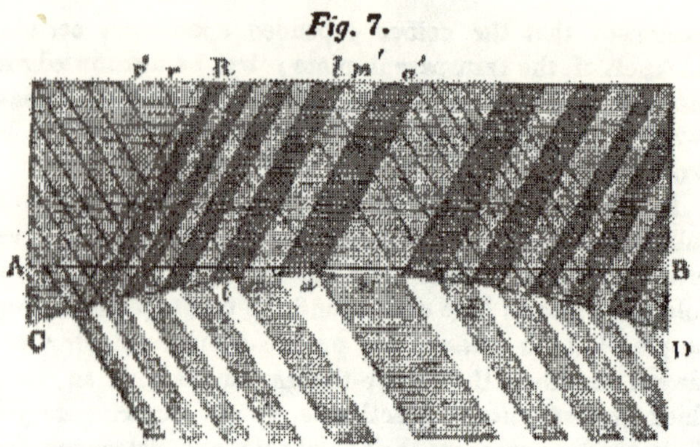

air is inconceivably thin, not a single ray of the pencil
RE is reflected. The light is wholly transmitted, and,
consequently, to an eye above E there will appear at E a
black spot. At *a*, where the plate of air is thicker, the
red light *ra* is reflected in the direction *aa'*; and as the
air has the same thickness in a circle round the point E,
the eye above E at *a* will see next the black spot E a ring
of red light. At *m*, where the thickness of the air is a
little greater than at *a*, the light *r' m* is all transmitted as
at E, and not a single ray suffers reflexion, so that to an
eye above E at *m'* there will be seen without the red ring
a a dark ring *m*. In like manner, at greater thicknesses
of the plate of air, there is a succession of *red* and dark
rings, diminishing in breadth, as shown in the diagram.

When the same experiment was repeated in *orange,
yellow, green, blue, indigo,* and *violet* light, the very same
phenomenon was observed; with this difference only, that
the rings were *largest* in *red* light, and *smallest* in *violet*
light, and had intermediate magnitudes in the intermediate
colours.

If the observer now places his eye below E, so as to see
the transmitted rays, he will observe a set of rings as

before, but they will have a bright spot in their centre at E, and the luminous rings will now correspond with those which were dark when seen by reflexion, as will be readily understood from inspecting the preceding diagram.

When the object-glasses are illuminated by *white* light, the *seven* systems of rings, formed by all the *seven* colours which compose white light, will now be seen at once. Had the rings in each colour been all of the same diameter, they would all have formed brilliant white rings, separated by dark intervals; but, as they have all different diameters, they will overlap one another, producing rings of various colours by their mixture. These colours, reckoning from the centre E, are as follows:—

1st Order. Black, blue, white, yellow, orange, red.
2d Order. Violet, blue, green, yellow, orange, red.
3rd Order. Purple, blue, green, yellow, red, bluish-red.
4th Order. Bluish-green, green, yellowish-green, red.
5th Order. Greenish-blue, red.
6th Order. Greenish-blue, red.

By accurate measurements, Sir Isaac found that the thicknesses of *air* at which the most luminous parts of the first rings were produced, were in parts of an inch $\frac{4}{178000}$, $\frac{5}{178000}$, $\frac{7}{178000}$, $\frac{8}{178000}$, $\frac{9}{178000}$, $\frac{11}{17800}$. If the medium or the substance of the thin plate is water, as in the case of the soap bubble, which produces beautiful colours according to its different degrees of thinness, the thicknesses at which the most luminous parts of the ring appear are produced at $\frac{1}{1336}$ of the thickness at which they are produced in air, and in the case of glass or mica at $\frac{1}{1525}$ of that thickness, the numbers 1·336, 1·525, expressing the ratio of the sines of the angles of incidence and refraction in the substances which produce the colours.

From the phenomena thus briefly described, Sir Isaac Newton deduced that ingenious, though hypothetical, property of light, called its *fits of easy reflexion and transmission.* This property consists in supposing that every particle of light from its first discharge from a luminous body possesses, at equally distant intervals, dispositions to be reflected from, and transmitted through, the surfaces of bodies upon which it is incident. Hence, if a particle of light reaches a reflecting surface of glass when in its *fit of easy reflexion,* or in its disposition to be reflected, it will yield more readily to the reflecting force of the surface; and, on the contrary, if it reaches the same surface while in a *fit of easy transmission,* or in a disposition to be transmitted, it will yield with more difficulty to the reflecting force. Sir Isaac did not venture to inquire into the cause of this property; but we may form a very intelligible idea of it by supposing, that the particles of light have two attractive and two repulsive poles at the extremities of two axes at right angles to each other, and that the particles revolve round their axes, and at equidistant intervals bring one or other of these axes into the line of direction in which the particle is moving. If the attractive axis is in the line of the direction in which the particle moves when it reaches the refracting surface, the particle will yield to the attractive force of the medium, and be refracted and transmitted; but if the repulsive axis is in the direction of the particle's motion when it reaches the surface, it will yield to the repulsive force of the medium, and be reflected from it.

The application of the theory of alternate fits of reflexion and transmission to explain the colours of thin plates is very simple. When the light falls upon the first surface AB (Fig. 7) of the plate of air between AB and CED, the rays that are in a fit of reflexion are reflected, and those that are in a fit of transmission are transmitted. Let us call F the length of a fit, or the distance through which the particle of light moves while it passes from the

state of being in a fit of reflexion to the state of being in
a fit of transmission. Now, as all the particles of light
transmitted through AB were in a state of easy transmission when they entered AB, it is obvious that, if the plate
of air at E is so thin as to be less than one-half of F, the
particles of light will still be in their disposition to be
transmitted, and consequently the light will be all transmitted, and none reflected at the curve surface at E.
When the plate becomes thicker towards *a*, so that its
thickness exceeds half of F, the light will not reach the
surface CE till it has come under its fit of reflexion, and,
consequently, at *a* the light will be all reflected and none
transmitted. As the thickness increases towards *m*, the
light will have come under its fit of transmission, and so
on, the light being reflected at *a*, *l*, and transmitted at E,
m. This will perhaps be still more easily understood from
Fig. 8, where we may suppose AEC to be a thin wedge

Fig. 8.

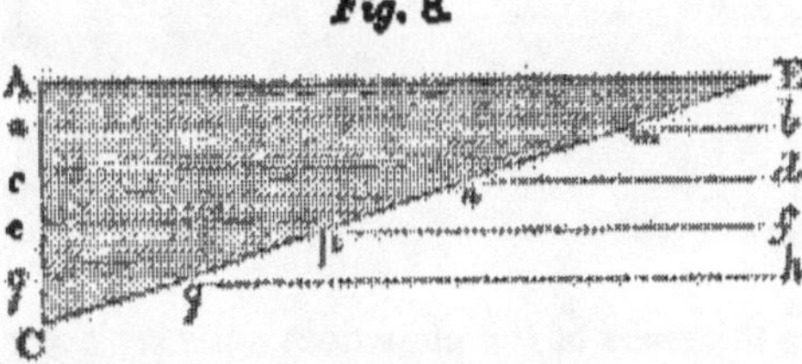

of glass or any other transparent body. When light is
incident on the first surface AE, all the particles of it
that are in a fit of easy reflexion will be reflected, and
all those in a fit of easy transmission will be transmitted.
As the fits of transmission all commence at AE, let the
first fit of transmission end when the particles of light
have reached *ab*, and the second when they have reached
ef; and let the fits of reflexion commence at *cd* and *gh*.
Then, as the fit of transmission continues from AE to *ab*,

all the light that falls upon the portion m E of the second
surface will be transmitted and none reflected, so that to
an eye above E the space m E will appear black. As the
fit of reflexion commences at ab, and continues to cd, all
the light which falls upon the portion nm will be reflected,
and none transmitted, and so on, the light being transmitted at m E and pn, and reflected at nm and qp. Hence
to an eye above E the wedge-shaped film, of which AEC
is a section, will be covered with parallel bands or fringes
of light separated by dark fringes of the same breadth,
and they will be all parallel to the thin edge of the plate,
a dark fringe corresponding to the thinnest edge. To an
eye placed below CE, similar fringes will be seen, but the
one corresponding to the thinnest edge m E will be
luminous.

Fig. 8.

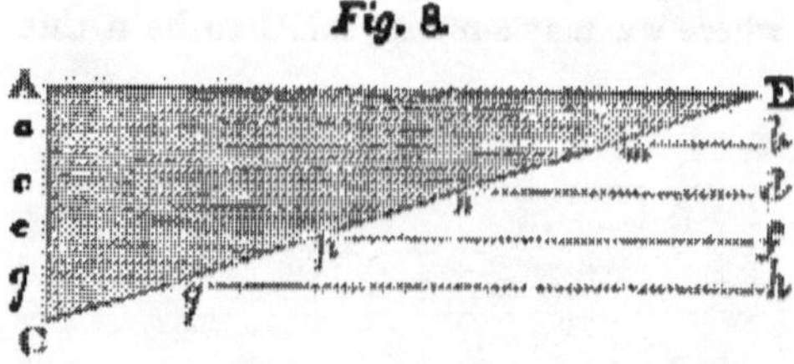

If the thickness of the plate does not vary according to
a regular law, as in Fig. 8, but if, like a film of blown
glass, it has numerous inequalities, then the alternate
fringes of light and darkness will vary with the thickness
of the film, and throughout the whole length of each
fringe the thickness of the film will be the same.

We have supposed in the preceding illustration that the
light employed is homogeneous. If it is white, then the
differently coloured fringes will form by their superposition
a system of fringes analogous to those seen between two
object-glasses, as already explained.

The same periodical colours which we have now described, as exhibited by thin plates, were discovered by Newton in thick plates, and he has explained them by means of the theory of fits; but it would lead us beyond the limits of a popular work like this to enter into any detail of his observations, or to give an account of the numerous and important additions which this branch of optics has received from the discoveries of succeeding authors.

CHAPTER VII.

*Newton's Theory of the Colours of Natural Bodies explained—
Objections to it stated—New Classification of Colours—
Outlines of a new Theory proposed.*

IF the objects of the material world had been illuminated with white light, all the particles of which possessed
the same degree of refrangibility, and were equally acted
upon by the bodies on which they fall, all nature would
have shone with a leaden hue, and all the combinations of
external objects, and all the features of the human countenance would have exhibited no other variety but that
which they possess in a pencil sketch, a China-ink drawing,
or a photographic picture. The rainbow itself would have
dwindled into a narrow arch of white light,—the stars
would have shone through a grey sky,—and the mantle of
a wintry twilight would have replaced the golden vesture
of the rising and setting sun. But He who has exhibited
such matchless skill in the organization of material bodies,
and such exquisite taste in the forms upon which they are
modelled, has superadded that ethereal beauty which en-
hances their more permanent qualities, and presents them
to us in the ever-varying colours of the spectrum. Without
this the foliage of vegetable life might have filled the eye
and fostered the fruit which it veils,—but the youthful
green of its spring would have been blended with the dying
yellow of its autumn. Without this the diamond, the ruby,
and the sapphire, might have displayed to science the nice
geometry of their forms, and yielded to the arts their

adamantine virtues; but they would never have sparkled
in the chaplet of beauty, or adorned the diadem of princes.
Without this the human face divine might have expressed
all the qualities of the mind, and beamed with all the
affections of the heart; but the "purple light of love,"*
would not have risen on the cheek, nor the hectic flush
been the herald of its decay.

The gay colouring with which the Almighty has decked
the pale marble of nature, is not the result of any quality
inherent in the coloured body, or in the particles by which
it may be tinged, but is merely a property of the light in
which they happen to be placed. Newton was the first
person who placed this great truth in the clearest evidence.
He found that all bodies, whatever were their peculiar
colours, exhibited these colours best in white light, or in
light which contained their peculiar colour. When they
are illuminated by homogeneous *red* light, they appear *red;*
by homogeneous *yellow* light, *yellow*, and so on; "their
colours being most brisk and vivid under the influence of
their own daylight colours." The leaf of a plant, for example, appeared *green* in the white light of day, because it
had the property of reflecting that light in greater abundance than any other. When it was placed in homogeneous
red light, it could no longer appear *green*, because there
was no green light to reflect; but it reflected a portion of
red light, because there was some red in the compound
green which it had the property of reflecting. If the leaf
had originally reflected a pure homogeneous green, un-
mixed with red, and reflected no white light (as all leaves
do) from its outer surface, it would have appeared quite
black in pure homogeneous red light, as this light does not
contain a single ray which the leaf is capable of reflecting.
Hence the colours of material bodies are owing to the

* The "lumen purpureum" of Virgil. "Scarlet glow" would be
more in accordance with our ideas of colour.—EDITOR.

property which they possess of stopping, or absorbing, certain rays of white light, while they reflect or transmit to the eye the rest of the rays of which white light is composed. The *green* leaf, for example, stops or absorbs the red, blue, and violet rays of the white light which falls upon it, and reflects and transmits only those which compose its peculiar green.

So far the Newtonian doctrine of colours is capable of rigid demonstration; but its author was not content with carrying it thus far. He sought to determine the manner in which particular rays are stopped, while others are reflected or transmitted; and the result of this profound inquiry was his theory of the colours of natural bodies, which was communicated to the Royal Society on the 10th of February, 1675. This theory is perhaps the loftiest of all his speculations; and though, as a physical generalization, it stands on a perishable basis, and must soon be swept away in the progress of optical discovery, it yet bears the deepest impress of the grasp of his powerful intellect.

The principles upon which this theory is founded are the following :—

1. Bodies that have the greatest refractive powers reflect the greatest quantity of light from their surfaces; and at the confines of equally refracting media there is no reflexion.

2. The least parts of almost all natural bodies are in some measure transparent.

3. Between the parts of opaque and coloured bodies are many spaces or pores, either empty or filled with media of other densities.

4. The parts of bodies and their interstices, or pores, must not be less than of some definite bigness, to render them coloured.

5. The transparent parts of bodies, according to their several sizes, reflect rays of one colour, and transmit those

of another on the same grounds that thin plates do reflect
or transmit these rays.

6. The parts of bodies on which their colours depend,
are denser than the medium which pervades their interstices.

7. The bigness of the component parts of natural bodies
may be conjectured by their colours.

Upon these principles Newton explains the origin of
transparency, *opacity*, and *colour*.

Transparency he considers as arising from the particles
and their intervals or pores being too small to cause reflexion at their common surfaces;* so that all the light
which enters transparent bodies passes through them without any portion of it being turned from its path by reflexion. If we could obtain, for example, a film of mica,
whose thickness does not exceed two-thirds of the mil-
lionth part of an inch, all the light which fell upon it
would pass through it, and none would be reflected. If
this film was then cut into fragments, a number of such
fragments would constitute a bundle, which would also
transmit all the light which fell upon it, and be perfectly
transparent.

Opacity in bodies arises, he thinks, from an opposite
cause, viz., when the parts of bodies are of such a size as
to be capable of reflecting the light which falls upon them,
in which case the light is "stopped or stifled" by the
multitude of reflexions.

The *colours* of natural bodies have, in the Newtonian
hypothesis, the same origin as the colours of thin plates,
their transparent particles, according to their several sizes,
reflecting rays of one colour, and transmitting those of
another. " For if a thinned or plated body which, being
of an uneven thickness, appears all over of one uniform
colour, should be slit into threads, or broken into fragments of the same thickness with the plate or film, every

* Optics, book ii., prop. iv.

thread or fragment should keep its colour, and consequently, a heap of such threads or fragments should constitute a mass or powder of the same colour which the plate exhibited before it was broken ; and the parts of all natural bodies being like so many fragments of a plate, must, on the same grounds, exhibit the same colour."

Such is the theory of the colours of natural bodies, stated as clearly and briefly as we can. It has been very generally admitted by philosophers, both of our own and of other countries, and has been recently illustrated and defended by a French philosopher of distinguished eminence. That this theory affords the true explanation of certain colours, or, to speak more correctly, that certain colours in natural bodies are the colours of thin plates, cannot be doubted ; but it will not be difficult to show that it is quite inapplicable to that great class of phenomena which may be considered as representing the colours of natural bodies.

The first objection to the Newtonian theory is the total absence of all reflected light from the particles of transparent coloured media, such as coloured gems, coloured glasses, and coloured fluids. This objection was urged long ago by Mr. Delaval, who placed coloured fluids on black grounds, and never could perceive the least trace of the reflected tints. I have repeated the experiment with every precaution, and with every variation that I could think of, and I consider it as an established fact, that in such coloured bodies the complementary reflected colour cannot be rendered visible. If the fluid, for example, be *red*, the *green* light from which the red has been separated ought to appear either directly by looking into the coloured mass, or ought to be recognised by its influence in modifying the light really reflected ; but as it cannot be seen, we must conclude that it has not been reflected, but has been destroyed by some other property of the coloured body.

A similar objection may be drawn from the disappearance of the transmitted complementary colour in the leaves of plants and petals of flowers. I have ascertained from numerous experiments, that the transmitted colour is almost invariably the same with the reflected colour, and that the same holds true with the coloured juices expressed from them. The complementary tints are never seen; and wherever there has been anything like an approximation to two tints, I have invariably found that it arose from there being two different coloured juices existing in different sides of the leaf.

In the phenomena of the light transmitted by coloured glasses, there are some peculiarities which we think demonstrate that their colours are not those of thin plates. The light, for example, transmitted through a particular kind of blue glass, has a blue colour of such a peculiar composition that there is no blue in any of the orders of colours in thin plates which has any resemblance to it. It is entirely destitute of the red rays which form the middle of the red space in the spectrum; so that the particles on which the colour depends must reflect the middle red rays, and transmit those on each side of it,—a property which cannot be deduced from the Newtonian doctrine.

The explanation of *opacity*, as arising from a multitude of reflexions, is liable to the same objection which we have urged against the explanation of colour. In order to appreciate its weight, we must distinguish opacity into two kinds, namely, the *opacity of whiteness* and the *opacity of blackness*. Those bodies which possess the power of reflexion in the highest degree, such as white metals, chalk, and plaster of Paris, never reflect more than one half of the light which falls upon them. The other half of the incident light is, according to Newton, lost by a multitude of reflexions. But how is it lost? Reflexion merely changes the direction of the particles of light, so that they must again emerge from the body, unless they are reflected into fixed returning orbits, which detain them for ever in a state of motion within the body. In the case of black opacity, such as that of coal, which reflects from its first surface only $\frac{1}{15}$th of the white light, the difficulty is still greater, and we cannot conceive how any system of interior reflexions could so completely stifle $\frac{14}{15}$ths of the whole incident light, without some of it returning to the eye in a visible form.

In determining the constitution of bodies that produces *transparency* and *blackness*, the Newtonian theory encounters a difficulty which its author has by no means surmounted. Transparency, as we have already seen, arises from the " particles and their interstices being too small to cause reflexions in their common surfaces;" that is, they must be "less than any of those which exhibit colours," or " less than is requisite to reflect the *white* and very faint *blue* of the first order. But *this is the very same constitution which produces blackness* by reflexion; and in order to explain the cause of blackness by transmission or black opacity, Newton is obliged to introduce a new principle.

" For the production of *black*," says he, " the corpuscles must be less than any of those which exhibit colours. For at all greater sizes there is too much light reflected to constitute this colour. But if they be supposed a little less than is requisite to reflect the white and very faint blue of the first order, they will reflect so very little light as to appear intensely black, *and yet may perhaps cariously refract* * it to and fro within themselves so long, until it

* In the same paragraph, when speaking of black bodies becoming hot, and burning sooner than others, he says that their " affect may proceed partly from the *multitude of refractions* in a little room, and partly from the easy commotion of so very small corpuscles."—*Optics*, part iii., prop. vii., p. 285.

happens to be stifled and lost, by which means they will appear black in all positions of the eye, without any transparency."

This very remarkable passage exhibits, in a striking manner, the perplexity in which our author was involved by the difficulties of his subject. As the particles which produce blackness by reflexion are necessarily so small as to exclude the existence of any reflective forces, he cannot ascribe the loss of the intromitted light, as he does in the case of white opacity, to a " multitude of reflexions;" and therefore he is compelled to have recourse to *refracting forces* to perform the same office. The reluctance with which he avails himself of this expedient, is well marked in the mode of expression which he adopts; and I am persuaded that when he wrote the above passage, he felt the full force of the objections to this hypothesis, which cannot fail to present themselves. As the size of the particles which produce blackness are intermediate between those which produce transparency and those which produce colour, approaching closely to the latter, it is difficult to conceive why *they* should refract the intromitted light, while the greater and smaller particles, and even those almost of the same size, should be destitute of that property. It is, besides, not easy to understand how a refraction can take place within bodies which shall stifle all the light, and prevent it from emerging. Nay, we may admit the existence of such refractions, and yet understand how, by a compensation in their direction, the refracted rays may all emerge from the opaque body.

The force of these objections is tacitly recognised in Pemberton's *View of Sir Isaac Newton's Philosophy;* and as Newton not only read and approved of that work, but even perused a great part of it along with its author (whose friendship and assistance in the preparation of the third edition of the *Principia* will be mentioned in the proper place), we may fairly consider the opinion there stated to be his own.

" For producing *black*, the particles ought to be smaller than for exhibiting any of the colours, viz., of a size answering to the thickness of the bubble, whereby reflecting little or no light, it appears colourless ; *but yet they must not be too small*, for that will make them transparent *through deficiency of reflexions* in the inward parts of the body, sufficient to stop the light from going through it; but they must be of a size *bordering upon that* disposed to reflect the faint blue of the first order, which affords an evident reason why blacks usually partake a little of that colour." In this passage all idea of refraction is abandoned, and that precise degree of size is assumed for the particles which leaves a small power of reflexion, which is deemed sufficient to prevent the body from becoming transparent; that is, sufficient to render it opaque or black.

The last objection which we shall state to this theory is one to which we attach great weight and, as it is founded on discoveries and views which have been published since the time of Newton, we venture to believe that, had he been aware of them, he would never have proposed the theory which we are considering.

When light falls upon a thin film such as AEC, (Fig. 8, p. 69,) so as to produce the colours of thin plates, it follows, from Sir Isaac Newton's theory of fits, that a portion of the light is, as usual, reflected at the first surface AE,[*] while the light which forms the coloured image is that which is reflected from the second surface EC, so that all the colours of thin plates are diluted with the white light

---

[*] When Newton speaks of bodies losing their reflecting power from their thinness, he means the reflecting power of their second surfaces, as is evident from the reason he assigns.—See Optics, part iii., prop. xiii., p. 257.

reflected from the first surface. Now, in the modern theory, which ascribes the colours of thin plates to the interference of the light reflected from the second surface EC, with the light reflected from the first surface AE, the resulting tint arises from the combination of these two pencils, and, consequently, there is no white light reflected from the surface AE. In like manner, when the thickness of the film is such that the two interfering pencils completely destroy one another, and produce black, there is not a ray of light reflected from the first surface. Here, then, we have a criterion for deciding between the theory of fits and the theory of interference; for if there is no white light reflected from the first surface AE, the theory of fits must be rejected. In a remarkable phenomenon of blackness arising from minute fibres, which I have had occasion to describe, there was no perceptible reflexion from the surface of the fibres; * and M. Fresnel describes an experiment made to determine the same point, and states the result of it to have been unequivocally in favour of the doctrine of interference.

In order to apply this important fact, let us take a piece of coal, one of the blackest and most opaque of all substances, and which does not reflect to the eye a single ray out of those which enter its substance. The size of its particles is so small that they are incapable of reflecting light. When a number of these particles are placed together, so as to form a surface, and other particles behind them, so as to form a solid, they will not acquire by this process the power of reflexion; and consequently, a piece of coal so composed should be destitute of the property of reflecting light from its first surface. But this is not the case,—light is abundantly reflected from the first surface of the coal, and consequently its elementary

* *Edinburgh Journal of Science*, No. I, p. 100.

particles must possess the same power. Hence the blackness of coal must be ascribed to some other cause than to the minuteness of its transparent atoms.

To transparent bodies this argument has a similar application. As their atoms are still less than those of black bodies, their inability to reflect light is still greater, and hence arises their transparency. But the particles forming the surface of such bodies do reflect light; and, therefore, their transparency must have another origin.

In the case of coloured bodies, too, the particles forming their surfaces reflect white light like those of all other bodies; so that these particles cannot produce colour on the same principles as those of thin plates. In many of those cases of colour which seem to depend upon the minuteness of the particles of the body, the reflexion of white light may nevertheless be observed; but this will be found to arise from a thin transparent film, behind which the colorific particles are placed.

Whatever answer may be given to these objections, we think it will be admitted by those who have studied the subject most profoundly, that a satisfactory theory of the colours of natural bodies is still a *desideratum* in science. How far we may be able to approach to it in the present state of optics, the reader will judge from the following views.

Colours may be arranged into seven classes, in each of which the colour has a different origin.

1. Transparent coloured fluids, such as the juices obtained from the coloured parts of plants, and coloured solutions, whether natural or artificial. Transparent coloured solids, such as coloured minerals, glasses, powders, and vegetable tissues—coloured gems—transparent coloured glasses—coloured powders—and the colours of the leaves and flowers of plants.

2. Oxidated films on metals—colours of Labrador feld-

spar—colours of precious and hydrophanous opal—of the feathers of birds—of the wings, &c., of insects—of the scales of fishes—of the *internal* films of mother of pearl and various shells—and of decomposed glass.

3. Superficial colours of mother of pearl, striated and grooved surfaces, which can be communicated by pressure to other surfaces.

4. Opalescences or colours dispersed from the particles of different solid and fluid and gaseous bodies, some of which are coloured and others colourless. These colours appear in ice, in water, in the atmosphere—in fluor spar and several glasses—in solutions of sulphate of quinine, &c., and in the juices of plants and several oils.

5. Colours at the surfaces of media of different dispersive powers, and in which the index of refraction is the same in each medium for certain rays, but different for all the rest.

6. Colours produced by heat, and during combustion.

7. The colours of metals.

The first two of these classes are the most important. The Newtonian theory appears to be strictly applicable to the phenomena of the *second* class; but those of the first class cannot, we conceive, be referred to the same cause.

The rays of solar light possess several remarkable physical properties: they heat—they illuminate—they promote chemical combination—they effect chemical decompositions—they impart magnetism to steel—they alter the colours of bodies—they communicate to plants and flowers their peculiar colours, and are in many cases necessary to the development of their characteristic qualities. It is impossible to admit for a moment that these varied effects are produced by a mere mechanical action, or that they arise from the agitation of the particles of bodies by the vibration of the ether which is con-

sidered to be the cause of light. Whatever be the difficulties which attach to the theory which supposes light
to consist of material particles, we are compelled, by its
properties, to admit that light acts as if it were material,
and that it enters into combinations with bodies, in order
to produce the effects which we have enumerated.

When a beam of light falls upon a body, and the whole
or part of that which enters its substance totally dis-
appears, we are entitled to say, that it is detained by some
power exercised by the particles of the body over the
particles of light. When this light is said to be lost by a
multitude of reflexions or refractions, the statement is not
only hypothetical, but it is a hypothesis incompatible with
optical principles. That the light detained within bodies
has been stopped by the attractive force of the particles
seems to be highly probable; and the mind will not feel
any repugnance to admit that the particles of all bodies,
whether solid, fluid, or aëriform, have a specific affinity
for the particles of light. Considering light, therefore, as
material, it is not difficult to comprehend how it should,
like other elementary substances, enter into combination
with bodies, and produce many chemical and physical
effects, but particularly the phenomena of transparency,
opacity, and colour.

In *transparent* colourless bodies, such as water and glass,
the intromitted light experiences a considerable loss, because a certain number of its particles are attracted and
detained by the atoms of the water or glass; and the light
which emerges is colourless, because the particles exercise
a proportional action over all the simple colours which
compose white light.

When the transparent body has any decided *colour*, such
as those enumerated in Class 1, then the particles of the
body have exercised a specific attraction over those rays
of white light which are complementary to those which

compose the colour of the transmitted light. If the transparent body, for example, is *red*, then its particles have detained the green rays which entered into the incident light, or certain other rays, which, with the red, are necessary to compose white light. In some artificial transparent bodies, like certain glasses, the particles will attract and detain rays of light of different colours, as may be seen by analysing the transmitted light with a prism, which will exhibit a spectrum deprived of all the rays which have been detained. In black bodies the particles exercise a powerful attraction over light, and detain all the intromitted rays.

When coloured bodies are opaque, so as to exhibit their colours principally by reflexion, the light which is reflected back to the observer has received its colour from transmission through part of the thickness of the body, or, what is the same thing, the colour reflected to the eye is complementary to that which has been detained by the particles of the body while the light is passing and repassing through a thickness terminated by the reflecting surfaces; and as only a part of this light is reflected, as in the case of leaves and flowers, the transmitted light must have the same colour as the reflected light.

When coloured bodies exhibit two different colours complementary to each other, the one seen by reflexion and the other by transmission, it is then highly probable that the colours are those of thin plates, though there are still other optical principles to which they may be referred. As the particles of bodies, and the medium which unites them, or as the different atoms of a compound body may have different dispersive powers, while they exercise the same refractive force over a particular part of the spectrum, the rays for which this compensation takes place will be transmitted, while part of the complementary light is reflected.* Or in case where the refractive and dispersive powers are the same, the reflective forces of the particles may vary according to a different law, so that at the separating surfaces either white or coloured light may be reflected.*

In these cases of colour where the reflected and the transmitted tints are not complementary, as in *leaf-gold*, where the former is *yellow* and the latter *green ;* in *leaf-silver*, where they are *white* and *blue;* and in certain pieces of fir-wood, where the reflected light is *whitish-yellow*, and the transmitted light a *brilliant homogeneous red ;* we may explain the separation of the colours either by the principles we have already laid down, or by the doctrine of thin plates. On the first principle, the colour of the reflected light, which is supposed to be the same as that of the transmitted light, will be modified by the law according to which the particles of the body attract different rays out of the beam of white light. In pitch, for example, the blue rays are first absorbed, so that at small thicknesses the transmitted light is a fine yellow, while, by the action of a greater thickness, the yellow itself is absorbed, and the transmitted light is a bright homogeneous red. Now in leaf-gold the transmitted colour of thinner films than we can obtain may be yellow, and, consequently, the light reflected from the first strata of interrupting faces will be yellow, and will determine the predominant tint of the reflected light. According to the Newtonian doctrine, on the other hand, it has been explained † by saying, " that the transmitted rays have traversed the whole thickness of the medium, and therefore undergo many more times the action of its atoms than those reflected, especially those near the first surface to which the brighter part of the reflected colour is due."

* See the *Phil. Trans.*, 1829, part I, p. 169.   † By Sir John Herschel.

The phenomena of the absorption of common and polarized light, which I have described in another place,* throw much light on the subject of coloured bodies. The relation of the absorbent action to the axes of double refraction, and, consequently, to the poles of the molecules of the crystal, shows how the particles of light attracted by the molecules of the body will vary, both in their nature and number, according to the direction in which they approach the molecules; and explains how the colour of a body may be changed, either temporarily or permanently, by heat, according as it produces a temporary or a permanent change in the relative position of the molecules. This is not the place to enlarge on this subject; but we may be permitted to apply the idea to the curious experiment of Thenard on phosphorus. When this substance is rendered pure by repeated distillation, it is transparent, and transmits yellow light; but when it is thrown in a melted state into cold water, it becomes jet black. When again melted, it resumes its original colour and transparency. According to the Newtonian theory, we must suppose that the atoms of the phosphorus have been diminished in size by sudden cooling,—an effect which it is not easy to comprehend; but, according to the preceding views, we may suppose that the atoms of the phosphorus have been forced by sudden cooling into relative positions quite different from those which they take when they slowly assume the solid state, and their poles of maximum attraction, in place of being turned to one another, are turned in different directions, and then allowed to exercise their full action in attracting the intromitted light, and detaining it wholly within the body.†

Before concluding this chapter, there is one topic pe-

* *Phil. Trans.*, 1819, p. 11.

† If this view of the matter be just, we should expect that the specific gravity of the black would exceed that of the yellow phosphorus.

culiarly deserving of our notice, namely, the change of
colour produced in bodies by continued exposure to light.
The general effect of light is to diminish or dilute the
colour of bodies, and in many cases to deprive them entirely of their colour.   Now, it is not easy to understand
how repeated undulations propagated through a body could
diminish the size of its particles, or how the same effect
could be produced by a multitude of reflexions from particle
to particle.   But if light is attracted by the particles of
bodies, and combines with them, it is easy to conceive,
that, when the molecules of a body have combined with a
great number of particles of a green colour, for example,
their power of combination with others will be diminished,
and, consequently, the number of particles of any colour
absorbed or detained must diminish with the time that the
body has been exposed to light; that is, these particles
must enter into the transmitted and reflected pencils, and
diminish the intensity of their colour.   If the body, for
example, absorbs red light, and transmits and reflects
green, then if the quantity of absorbed red light is diminished, it will enter into the reflected and transmitted
pencils, and, forming white light by its mixture with a
portion of the green rays, will actually dilute them in the
same manner as if a portion of white light had been added.*

* In a note here appended, Sir D. Brewster remarks that " since the
two preceding chapters were written, I have had occasion to confirm and
extend the views which they contain by many new experiments."   And
he refers in the larger edition to the interesting discoveries of Professor
Stokes of Cambridge, relating to some remarkable colours produced by
internal dispersion of light, which " must be reflected in cases of ordinary
opalescence, from the faces of minute pores in solids, or from particles of
different densities, disseminated through solids, or suspended in fluids."—
EDITOR.

CHAPTER VIII.

Newton's Discoveries respecting the Inflexion or Diffraction of Light—Previous discoveries of Grimaldi and Dr. Hooke— Labours of succeeding Philosophers—Law of Interference of Dr. Young—Fresnel's discoveries—The Undulatory Theory generally.

AMONG the optical discoveries of Newton, those which he made on the *Inflexion of Light* hold a high place. They were first published in his Treatise on Optics in 1704, but we have not been able to ascertain at what period they were made. In the preface to that work, Sir Isaac informs us that the third book, which contains his experiments on inflexion, "was put together out of scattered papers," and he adds at the end of his observations, that he "designed to repeat most of them with more care and exactness, and to make some new ones for determining the manner how the rays of light are bent in their passage by bodies, for making the fringes of colours with the dark lines between them. But we were then interrupted, and cannot now think of taking these things into consideration."

The phenomena of the inflexion of light were first discovered by Francis Maria Grimaldi, a learned Jesuit, who has described them in a posthumous work published in 1665, two years after his death.*

Having admitted a beam of the sun's light through a

* *Physico-Mathesis de Lumine, Coloribus, et Iride, aliisque annexis.* Bonn. 1665.

small pin-hole in a piece of lead or card into a dark chamber, he found that the light diverged from this aperture in the form of a cone, and that the shadows of all bodies placed in this light were not only larger than might have been expected, but were surrounded with three coloured fringes, the nearest being the widest, and the most remote the narrowest. In strong light he discovered analogous fringes within the shadows of bodies, which increased in number with the breadth of the body, and became more distinct when the shadow was received obliquely and at a greater distance. When two small apertures or pin-holes were placed so near each other that the cones of light formed by each of them intersected one another, Grimaldi observed, that a spot common to the circumference of each, or, which is the same thing, illuminated by rays from each cone, was darker than the same spot when illuminated by either of the cones separately; and he announces this remarkable fact in the following paradoxical proposition, "*that a body actually illuminated may become more dark by adding a light to that which it already receives.*" Grimaldi discovered also the beautiful phenomenon of the crested or curved fringes exhibited within the shadow of the rectangular termination of bodies.

Without knowing what had been done by the Italian philosopher, our countryman, Dr. Robert Hooke, had been diligently occupied with the same subject. In 1672, he communicated his first observations to the Royal Society, and he then spoke of his paper as " containing the discovery of a new property of light not mentioned by any optical writers before him." In a valuable memoir read by him on the 10th of March, 1674, he not only described the leading phenomena of the inflexion, (or the deflexion of light, as he called it), but he distinctly announced the *doctrine of interference,* which has performed so great a part in the subsequent history of optics.*

Such was the state of the subject when Newton directed to it his powers of acute and accurate observation. His attention, however, was turned only to the enlargement of the shadow of inflecting bodies, and to the three fringes adjacent to it. He was therefore led to take exact measures of the diameter of the shadow of a human hair, and of the breadth of the fringes at different distances behind it, and repeating these observations with light of different colours, he obtained two new and remarkable results :—(1.) That these breadths were not proportional to the distances from the hair at which they were measured; and (2.) That the fringes made in homogeneous *red* light were *red*, and the largest; that those made in *violet* light were *violet*, and the smallest; and that those made in *green* light were *green*, and of an intermediate size; the rays which formed the red fringe passing by the hair at a greater distance than those which formed the violet.

As already mentioned, when Newton made the preceding observations, he intended to repeat most of them with more care, and to make "some new ones to determine the manner how the rays of light are bent in their passage by bodies;" but having been then interrupted, he could not think of resuming the inquiry.

It is very difficult to ascertain his real views on the subject of *inflexion*. In his Discourse read in 1675, he ascribes it to the variable density of the ether within and without the inflecting body, thus regarding it as a new

* This doctrine is thus announced. 1. That the same rays of light falling upon the same point of an object will turn into all sorts of colours by the various inclination of the object. 2. That colours begin to appear when two pulses of light are blended so well and so near together that the sense takes them for one.

species of refraction; and in his letter to Robert Boyle in 1679, he takes the same view of the subject, and considers the several colours of the fringes as produced "by that refraction." Pursuing the same idea, he asserts in the Scholium to the 96th Prop. of the first book of the *Principia*, that the rays of light, in passing near bodies, are bent round them as if by attraction; that the rays which pass nearest them are most bent, as if they were most attracted; that those which pass at a greater distance are less bent; and that those which pass at still greater distances are bent in an opposite direction.

In this remarkable passage Newton introduces, for the first time, the idea of a force bending the rays outwards; or of an *inflecting* force bending the rays inwards, accompanied with a *deflecting* force bending them outwards. This opinion, however, he subsequently abandoned; for, in the third book of his *Optics*, he refers all the phenomena to a force which "bends the rays *not towards*, but *from* the shadow;" and he distinctly asserts, "that light *is never known to follow crooked passages, nor to bend into the shadow*."

These erroneous opinions, now wholly exploded, arose from Newton's having never observed the *internal fringes*, or those seen within the shadow. Grimaldi had described them minutely in his work, and as they have been seen by almost every philosopher, it is not easy to explain how they should have escaped the notice of two such careful observers as Hooke and Newton. Without this cardinal fact, our author stumbled in his path, and was misled into the erroneous propositions that bodies act upon light at a distance; that this action bends its rays with a force diminishing with the distance; and that rays which differ in refrangibility differ also in flexibility. Nor was he nearer the truth when he conjectured in his third query that the rays of light, in passing by the edges of bodies, may be bent several times backwards and forwards with a motion like that of an eel, and that the three fringes of coloured light may arise from three such bendings.

A subject which had thus baffled the sagacity of Newton, was not likely to unfold its mysteries to ordinary observers. The experiments of Grimaldi and Newton were repeated by various philosophers in various lands. Observations better made, and measures more accurately taken, were continually accumulating. But it was not until more than a century had passed away that experiments were made which led to a different view of the whole subject, and effectually dissipated the idea that the light which passed by the edges of bodies was inflected by any refracting agent, or by any action whatever of the bodies themselves.

During two sets of experiments which I made on the inflexion of light, the first in 1798, and the second in 1812 and 1813, I was desirous of examining the influence of density and refractive power over the fringes produced by inflexion. I compared the fringes formed by gold-leaf with those formed by masses of gold, and those produced by films which gave the colours of thin plates with those formed by masses of the same substance. I examined the influence of platinum, diamond, and cork in inflecting light, the effect of non-reflecting grooves and spaces in polished metals, and of cylinders of glass immersed in a mixture of oil of cassia and oil of olives of the same refractive power; and, as the fringes had the same magnitude and character under all these circumstances, I concluded that they were not produced by any force inherent in the bodies themselves, but arose from a property of the light, which always showed itself when light was stopped in its progress.

It is to Dr. Thomas Young, however, that we owe the master fact which enabled us to unveil the mysteries of

diffraction, and to account for a great variety of hitherto
unexplained phenomena. In studying the internal fringes
and the crested ones discovered by Grimaldi, he found
that by intercepting the rays which passed by one side of
the diffracting body, the internal and the crested fringes
completely disappeared; and hence he concluded that the
fringes were produced by the joint action, or by the
interference, of the two portions of light which passed on
each side of the diffracting body. In order to account
for the coloured fringes without the shadow, Dr. Young
conceived that the rays which pass near the edge of the
hair interfere with others, which he supposed might be reflected after falling very obliquely upon its edge,—a supposition which, if correct, would certainly produce fringes
very similar to those actually observed.

In pursuing the researches so successfully begun by
Dr. Young, M. Fresnel had the good fortune to explain
all the phenomena of inflexion by means of the undulatory
doctrine combined with the principle of interference. In
place of transmitting the light through a small aperture,
he caused it to diverge from the focus of a deep convex
lens, and, instead of receiving the shadow and its fringes
upon a smooth white surface, as was done by Newton, he
viewed them directly with his eye through a lens placed
behind the shadow; and by means of a microscope he
was able to measure the dimensions of the fringes with
the greatest exactness. By this mode of observation he
made the remarkable discovery, that the inflexion of the
light *depended on the distance of the inflecting body from
the aperture or from the focus of divergence,* the fringes
being observed to dilate as the body approached that focus,
and to contract as it receded from it, their relative dis-

* This effect is so great, that, at the distance of *four* inches from the
point of divergence, the angular inflexion of the *red* rays of the first
fringe is 12′ 6″, while at the distance of about twenty feet it is only 8′ 55″,

tances from each other and from the margin of the shadow continuing invariable. In attempting to account for the formation of the exterior fringes, M. Fresnel found it necessary to reject the supposition of Dr. Young, that they were owing to light reflected from the edge of the body. He not only ascertained that the real place of the fringe was the $\frac{1}{100}$dth of a millimetre different from what it should be on that supposition, but he found that the fringes preserved the same intensity of light, whether the inflecting body had a round or a sharp edge, and even when the edge was such as not to afford sufficient light for their production. From this difficulty the undulatory theory speedily released him, and he was led by its indications to consider the exterior fringes as produced by an infinite number of elementary waves of light, emanating from a primitive wave when partly interrupted by an opaque body.

In all the experiments on inflexion and diffraction made by Newton and Fresnel, the fringes were viewed either on paper, or in the focus of a lens when the rays had actually interfered and produced the coloured fringes. The fringes thus seen may be called *positive*, because they are formed in space and out of the eye, on the retina of which they are afterwards delineated; but there is another form of these fringes which I have examined, and which may be called *negative*, because they are not brought to a positive focus in space, or do not interfere till they reach the retina. In order to see these fringes, place the lens behind the diffracting body, so as to see the *positive* fringes, and then move it forward till these fringes disappear. The diffracting edge will now be in the anterior focus of the lens. If we advance the lens towards the diffracting body, the *negative* fringes will appear, and will increase in size till the lens touches the body, when they will have the same magnitude as the *positive* fringes have

when the lens is placed behind the body, at the distance
of twice its focal length.

If we wish to see the fringes larger, we must use a lens
with a longer focus; and when it is placed in contact
with the diffracting body, the fringes will in every case be
the same as the positive ones seen by the same lens placed
behind the body twice its focal length. If the diffracting
body is included in a fluid lens, or even placed *in front of
the lens*, the negative fringes will be seen. In producing
the negative fringes, the interfering rays are those which
virtually radiate from the anterior focus of the lens, and
which, being refracted into parallel directions, enter the
eye, and interfere on the retina; and in consequence of
their not interfering till they enter the eye, they are much
more distinct than the positive fringes.*

Whatever opinion we may form of the undulatory
theory in its physical aspect, the explanation which it
affords of a vast variety of optical phenomena, entitles it
to the highest consideration. With few exceptions, this
theory has been shown to give a satisfactory explanation
of the leading phenomena of diffraction, while the Newtonian or corpuscular hypothesis has not even ventured to
suggest one which is probable.

* Sir D. Brewster here refers, in his larger edition, to some interesting
experiments on inflexion made by Lord Brougham in Provence, the
results of which were published in 1850. But perhaps, in a memoir of
Sir Isaac Newton, enough has been said on the subject. The undulatory
theory of light has now been established on too firm a basis ever to be
shaken; the territory for further scientific acquisition is the nature and
form of the undulations or vibrations.—EDITOR.

CHAPTER IX.

Miscellaneous Optical Researches of Newton—His Experiments on Refraction—His Conjecture respecting the Inflammability of the Diamond—His Law of Double Refraction— His Observations on the Polarization of Light—Newton's Theory of Light—His " Optics."

BEFORE concluding our account of Newton's optical discoveries, it is necessary to notice some of his minor researches, which, though of inferior importance in the science of light, have either exercised an influence over the progress of discovery, or been associated with the history of other branches of knowledge.

One of the most curious of these inquiries related to the connexion between the refractive powers and the chemical composition of bodies. Having measured the refractive powers and densities of twenty-two substances, he found that the forces which reflect and refract light are very nearly proportional to the densities of the same bodies. In this law, however, he noticed a remarkable exception in the case of unctuous and sulphureous bodies, such as camphor, olive oil, linseed oil, spirit of turpentine, and *diamond*, which have their refractive powers two or three times greater in respect of their densities than the other substances in the list, while among themselves their refractive powers are proportional to their densities, without any considerable variation. Hence he concluded that diamond "is an unctuous substance congulated,"— a sagacious prediction, which has been verified in the discoveries of modern chemistry. The connexion between a high degree of inflammability, and a great refracting force, has been still more strongly established by the high refractive power which I detected in phosphorus, and which was discovered in hydrogen by MM. Biot and Arago.

There is no part of the optical labours of Newton which is less satisfactory than that which relates to the *double refraction* of light. In 1690, Huygens published his admirable treatise on light, in which he has given the law of double refraction in calcareous spar, as deduced from his theory of light, and as confirmed by direct experiment. Viewing it probably as only a theoretical deduction, Newton seems to have regarded it as incorrect; and though he gave Huygens the credit of describing the phenomena more exactly than Bartholinus, yet, without assigning any reason, he rejected the law of the Dutch philosopher, and substituted another in its place. These observations of our author form the subject of the twenty-fifth and twenty-sixth queries at the end of his Optics, which was published fourteen years after the appearance of Huygens's work. The law adopted by Newton is not accompanied with any of the experiments from which it was deduced, though it was stated by him as an undoubted truth. But it has been since shown to be erroneous; first, by the Abbé Hauy, and has been rejected by all succeeding philosophers.

In his speculations respecting the successive disappearance and reappearance of two of the four images which are formed when a luminous object is viewed through two rhombs of calcareous spar, one of which is made to revolve upon the other, Newton was more successful, though he omitted to give to Huygens the credit of having discovered these curious phenomena. He concluded, from his observations of them, that every ray of light has two opposite sides originally endued with the property on which the unusual refraction depends, and other two opposite sides not endued with that property; and he suggested it as a subject for future inquiry, whether there are not more properties of light by which the sides of the rays differ, and are distinguished from one another. This is the first occasion on which the idea of a *polarity* in the rays of light has been suggested.[*]

More than a century elapsed after Newton's observations before the double refraction and polarization of light in Iceland spar, and other bodies were reduced to regular laws. In 1810, Malus announced to the Institute of France the remarkable discovery that a ray of light reflected at a particular angle was polarized like one of the pencils formed by Iceland spar; that is, exhibited the same properties in its four sides or quarters which are exhibited in one of the pencils of Iceland spar; and the result of this fine discovery has been the establishment of a new branch of Physical Optics, which possesses the highest interest, not only from the beauty of its laws and the splendour of its phenomena, but from the new power with which it arms the philosopher in detecting organic or inorganic structures, which defy the scrutiny of the eye and the microscope.

Whilst on the subject of optics, mention should be made[†] of the reflecting sextant invented by Newton, for observing the Moon's distance from the fixed stars at sea. The description of this instrument was communicated to Dr. Halley in the year 1700; but, either from having mislaid the manuscript, or from attaching no value to the invention, he never communicated it to the Royal Society, and it remained among his papers till after his

---

[*] See the twenty-ninth Query at the end of his Optics, where the sides of a ray are compared with the poles of a magnet.

[†] It will be described farther on.—Editor.

death, and was read to the Society on the 28th of October, 1742. From the description it is obviously the very same invention as that which Hadley produced in 1731, and which, under the name of Hadley's Quadrant, has been of so great service in navigation. But though the first merit of this invention is thus transferred to Newton, we must not omit to state that the germ of it, and something more, had been previously published by Hooke.

The most important of the optical discoveries of Newton, of which we have given a general history, were communicated to the Royal Society in detached papers; but the disputes in which they had involved their author, made him hesitate about the publication of his other discoveries. Although he had drawn up a connected view of his labours under the title of "Opticks, or Treatise on the Reflexions, Refractions, Inflexions, and Colours of Light," yet he resolved not to publish this work during the life of Hooke, by whose rival jealousy his tranquillity had been so frequently interrupted. Hooke, however, died in 1702, and the Optics of Newton appeared in English in 1704. Dr. Samuel Clarke proposed a Latin edition of it, which appeared in 1706; and he was generously presented by Sir Isaac with £500, (or £100 for each of his five children,) as a token of the approbation and gratitude of the author. Both the English and the Latin editions have been frequently reprinted both in England and on the Continent,* and perhaps there never was a work of profound science more widely circulated.

* The English edition was reprinted at London in 1714, 1721, and 1730, and the Latin one at London in 1706, 1710, 1721, 1726, at Lausanne in 1740, and at Padua in 1778.

CHAPTER X.

*Astronomical Discoveries of Newton—Necessity of combined
Exertion to the Completion of great Discoveries—Sketch of
the History of Astronomy previous to the Time of Newton
—Copernicus, 1472–1543 — Tycho Brahe, 1546–1601—
Kepler, 1571–1630—Galileo, 1564–1642.*

FROM the optical labours of Newton we now proceed
to the history of his astronomical discoveries, those transcendent deductions of human reason by which he has
secured to himself an immortal name, and vindicated the
intellectual dignity of his species. Pre-eminent as his
triumphs have been, it would be unjust to affirm that they
were achieved by his single arm. The torch of many a
preceding age had thrown its light into the strongholds of
the material universe, and the grasp of many a powerful
hand had pulled down the most impregnable of its defences.
An alliance, indeed, of many kindred spirits had been
long struggling in this great cause, and Newton was but
the leader of their mighty phalanx, the director of their
combined genius, the general who won the victory, and
therefore wears its laurels.

The history of science presents us with no example of
an individual mind throwing itself far in advance of its
contemporaries. It is only in the career of crime and
ambition that reckless man takes the start of his species;
and, uncurbed by moral and religious ties, represses the
claims of truth and justice, and founds an unholy empire
upon the ruins of ancient and venerable institutions. The
achievements of intellectual power, though frequently

begun by one mind and completed by another, have ever been the results of united labour. Slow in their growth, they gradually approximate to a more perfect condition. The variety in the objects and phenomena of nature summons to research a variety of intellectual gifts: observation collects her materials, and patiently plies her humble vocation: experiment, with her quick eye and ready hand, developes new facts: the lofty powers of analysis and combination generalize insulated results and establish physical laws; and in the ordeal of contending schools and rival inquirers truth is finally purified from error. How different is it with those systems which the imagination rears, those theories of wild import which are directed against the liberties, the consciences, and the hopes of man! The fatal poison tree distils its virus in the spring as well as in the autumn of its growth, but the fruit which sustains life must have its bud prepared before the approach of winter, its blossom expanded in the spring, and its juices elaborated by the light and heat of both a summer and an autumnal sun.

In the century which preceded the birth of Newton, the science of astronomy advanced with the most rapid pace. Emerging from the darkness of the middle ages, the human mind seemed to rejoice in its new-born strength, and to apply itself with elastic vigour to unfold the mechanism of the heavens. The labours of Hipparchus and Ptolemy had indeed furnished many important epochs and supplied many valuable data; but the cumbrous appendages of cycles and epicycles with which they explained the stations and retrogradations of the planets, and the vulgar prejudices which a false interpretation of Scripture had excited against a belief in the motion of the earth, rendered it difficult even for great minds to escape from the trammels of authority, and appeal to the simplicity of nature.

In the thirteenth century, the noble-minded Alphonso X, sovereign of Castile, published at a great expense new astronomical tables, computed by the most distinguished professors in the Moorish universities; and as if he had obtained a glimpse of a simpler arrangement, he is said to have denounced the rude mechanism of epicycles in language less reverent in its expression than in its truth, declaring that if the heavens were thus constituted, he could have given the Deity good advice! But neither he nor the astronomers whom he so liberally protected seem to have established a better system, and it was left to Copernicus to enjoy the dignity of being the restorer of astronomy.

This great man, of Bohemian origin, but a native of Thorn in Prussia, where he was born on the 19th of January, 1472, followed at first his father's profession, and began his career as a Doctor of Medicine, but an accidental attendance on the mathematical lectures of Brudzevius excited a love for astronomy, which became the leading passion of his life. Quitting a profession uncongenial to such pursuits, he went to Bologna to study astronomy under Dominic Maria, and after having enjoyed the friendship and instruction of that able philosopher, he established himself at Rome in the humble position of a teacher of mathematics. Here he made numerous astronomical observations which served him as the basis of future researches; but an event soon occurred, which, though it interrupted for a while his important studies, placed him in a situation for pursuing them with new zeal. The death of one of the canons enabled his uncle, who was Bishop of Ermeland, to appoint him to a canonry in the chapter of Frauenberg, where, in a house situated on the brow of a mountain, he continued, in peaceful seclusion, to carry on his astronomical observations. During his residence at Rome his talents had been so well appreciated,

that the Bishop of Fossombrona, who presided over the
council for reforming the calendar, solicited the aid of
Copernicus in this desirable undertaking. At first he
entered warmly into the views of the council, and charged
himself with the determination of the length of the year
and of the month, and of the other motions of the sun and
moon that seemed to be required; but he found the task
too irksome, and probably felt that it would interfere with
those interesting discoveries which had already begun to
dawn upon his mind.

Copernicus is said to have commenced his inquiries by
a historical examination of the opinions of ancient authors
on the system of the universe; but it is more likely that
he sought for the authority of their great names to countenance his own peculiar views, and that he was desirous
to present his new theory as one that he had received,
rather than as one which he had invented. His mind had
been long imbued with the idea, that simplicity and
harmony should characterize the arrangements of the
planetary system; and in the complication and disorder
which reigned in the hypothesis of Ptolemy, he saw insuperable objections to its being regarded as a representation of nature. In the opinions of the Egyptian sages, in
those of Pythagoras, Philolaus, Aristarchus, and Nicetas,
he recognised his own earliest conviction that the earth
was not the centre of the universe; but he appears to have
considered it as still possible that our globe might perform
some function in the system more important than that of
the other planets; and his attention was much occupied
with the speculation of Martianus Capella, who placed the
sun between Mars and the Moon, and made Mercury and
Venus revolve round him as a centre; and with the system
of Apollonius Pergæus, who made all the planets revolve
round the sun, while the sun and moon were carried round
the earth in the centre of the universe. The examination,

however, of these hypotheses gradually dispelled the difficulties with which the subject was beset; and after thirty-
six years of intense study, in which the labours of the
observer and the calculations of the mathematician were
combined with the sagacity of the philosopher, he was
permitted to develop the true system of the heavens. The
sun he considered as immovable in the centre of the system,
while the earth revolved annually round him between
the orbits of Venus and Mars, producing by its rotation
about its axis in twenty-four hours all the diurnal phenomena of the celestial sphere. The precession of the
equinoxes was thus referred to a slight motion of the
earth's axis, and the stations and retrogradations of the
planets were the necessary consequence of their own
motions combined with that of the earth about the sun.
These remarkable views were supported by numerous astronomical observations; and in 1530 Copernicus brought
to a close his immortal work on the Revolutions of the
Heavenly Bodies.

But while we admire the genius which triumphed over
so many difficulties, we cannot fail to commend the extraordinary prudence with which he ushered his new system
into the world. Aware of the prejudices, and even of the
hostility, with which such a system would be received, he
resolved neither to startle the one nor provoke the other.
He allowed his opinions to circulate in the slow current of
personal communication. The points of opposition which
they presented to established doctrines were gradually worn
down, and they insinuated themselves into reception
among the ecclesiastical circles by the very reluctance of
their author to bring them into notice. In the year 1536,
Cardinal Schonberg, Bishop of Capua, and Gyse, Bishop
of Culm, exerted all their influence to induce Copernicus
to lay his system before the world; but he resisted their
solicitations; and it was not till 1539 that an accidental

circumstance contributed to alter his resolution. George Rheticus, Professor of Mathematics at Wirtemberg, having heard of the labours of Copernicus, resigned his chair, and repaired to Frauenberg to make himself master of his discoveries. This zealous disciple prevailed upon his master to permit the publication of his system; and they seem to have arranged a plan for giving it to the world without alarming the vigilance of the church, or startling the prejudices of individuals. Under the disguise of a student of mathematics, Rheticus published in 1540 an account of the manuscript volume of Copernicus. This pamphlet was received without any disapprobation, and its author was encouraged to reprint it at Basle, in 1541, with his own name. The success of these publications, and the flattering manner in which the new astronomy was received by several able writers, induced Copernicus to place his MS. in the hands of Rheticus. It was accordingly printed at the expense of Cardinal Schonberg, and appeared at Nuremberg in 1543, under the title, "*On the Revolutions of the Celestial Bodies.*" Its illustrious author, however, did not live to peruse it. A complete copy was handed to him on his dying day, and he saw and touched it a few hours before he expired. This great work was dedicated to the holy Pontiff, in order, as Copernicus himself says, that the authority of the head of the church might silence the calumnies of individuals who had attacked his views by arguments drawn from religion. Thus introduced, the Copernican system met with no ecclesiastical opposition, and gradually made its way in spite of the ignorance and prejudices of the age.

Among the astronomers who provided the materials of the Newtonian philosophy, the name of Tycho Brahe merits a conspicuous place. Descended from an ancient Swedish family, he was born at Knudstrup, in Norway, on the 14th December, 1546, three years after the death of Copernicus. The great eclipse of the sun which happened on the 26th of August, 1560, while he was at the University of Copenhagen, attracted his notice; and when he found that all its phenomena had been accurately predicted, he was seized with the most irresistible passion to acquire the knowledge of a science so infallible in its results. Destined for the profession of the law, his friends discouraged the pursuit which now engrossed his thoughts, and such were the reproaches, and even persecutions to which he was exposed, that he quitted his country with the design of travelling through Germany. During a visit to Augsburg, he inspired the burgomaster of the city, Peter Hainzell (at whose house he resided), with such a love of astronomy that he erected an excellent observatory at his own expense, and thus enabled his youthful instructor to commence that splendid career of observation, which has placed him in the first rank of practical astronomers.

Upon his return to Copenhagen in 1570, he was received with every mark of respect. The king invited him to court, and persons of all ranks harassed him with their attentions. At Herritzvold, near his native place, the house of his maternal uncle afforded him a retreat from the gaieties of the capital, and he was there offered every accommodation for the prosecution of his astronomical studies. Here, however, the passion of love and the pursuits of alchemy distracted his thoughts; but he found the peasant girl of whom he became enamoured of easier attainment than the philosopher's stone. His noble relatives were deeply offended with the marriage, and it required all the influence of the king to allay the quarrel which it occasioned. In the tranquillity of domestic happiness, Tycho resumed his study of the heavens; and, in 1572-3, he enjoyed the singular good fortune of observing, through all its variations, the new star in Cassiopeia, which appeared with such extraordinary splendour as to be visible in the day-time, and gradually disappeared in the following year.

Dissatisfied with his residence in Denmark, Tycho resolved to settle in some distant country; and having gone as far as Venice in search of a suitable residence, he at last fixed upon Basle, in Switzerland. The King of Denmark, however, had learned his intention from the Prince of Hesse; and when Tycho returned to Copenhagen to remove his family and his instruments, his sovereign announced to him his resolution to detain him in his kingdom. He presented him with the canonry of Roschild, with an income of 2000 crowns per annum. To this he added a pension of 1000 crowns; and he promised to give him the Island of Huen, with a complete observatory erected under his own eye. This generous offer was instantly accepted. The celebrated observatory of Uraniburg was established at the expense of about £20,000; and in this magnificent retreat Tycho continued for twenty-one years to enrich astronomy with the most valuable observations. Admiring disciples crowded to this sanctuary of the sciences to acquire the knowledge of the heavens; and kings* and princes felt themselves

---

* When James I. went to Copenhagen in 1590, to conclude his marriage with the Princess Anne of Denmark, he spent eight days under the roof of Tycho at Uraniburg. As a token of his gratitude, he composed a set of Latin verses in honour of the astronomer, and left him a magnificent present at his departure. He gave him also his royal licence for the publication of his works in England, and accompanied it with the following complimentary letter:

"Nor am I acquainted with these things on the relation of others, or from a mere perusal of your works, but I have seen them with my own eyes, and heard them with my own ears, in your residence at Uraniburg, during the various learned and agreeable conversations which I there held with you, which even now affect my mind to such a degree, that it is difficult to decide whether I recollect them with greater pleasure or admiration."

hononred by becoming the guests of the great astronomer of the age.

One of the principal discoveries of Tycho was that of the inequality of the moon's motion, called the Variation. He detected also the annual equation which affects the place of her apogee and nodes, and he determined the greatest and the least inclination of the lunar orbit. His observations of the planets were numerous and precise, and formed the data of the subsequent generalizations in astronomy. Though thus skilful in the observation of phenomena, his mind was but little suited to investigate their cause, and it was probably owing to this defect that he rejected the system of Copernicus. The vanity of giving his own name to another system was not likely to actuate a mind such as his; and it is more probable that he was led to adopt the immobility of the earth, and to make the sun, with all his attendant planets, circulate round it, from the great difficulty which still presented itself by comparing the apparent diameter of the stars with the annual parallax of the earth's orbit.

The death of Frederick in 1588 proved a severe calamity to Tycho, and to the science which he cultivated. During the first years of the minority of Christian IV., the regency continued the royal patronage to the observatory of Uraniburg; and in 1592, the young king paid a visit of some days to Tycho, and left him a gold chain in token of his favour. The astronomer, however, had made himself enemies at court, and the envy of his high reputation had probably added fresh malignity to the irritation of personal feelings. Under the ministry of Walchendorf, a name for ever odious to science, Tycho's pension was stopped;—he was, in 1597, deprived of the canonry of Roschild, and was thus forced, with his wife and children, to seek an asylum in a foreign land. His friend, Henry Rantzau of Wansbeck, under whose roof he found a hospitable shelter, was fortunately acquainted with the Emperor Rodolph II., who, to his love of science, added a passion for alchemy and astrology. The reputation of Tycho having already reached the imperial ear, the recommendation of Rantzau was scarcely necessary to insure him his warmest friendship. Invited by the emperor, he repaired in 1599 to Prague, where he met with the kindest reception. A pension of three thousand crowns was immediately settled upon him, and a commodious observatory erected for his use in the vicinity of that city. Here the exiled astronomer renewed with delight his interrupted labours, and the gratitude which he cherished for the royal favour increased the satisfaction which he felt in having so unexpectedly found a resting-place for approaching age. These prospects of better days were enhanced by the good fortune of receiving two such men as Kepler and Longomontanus for his pupils; but the fallacy of human anticipation was here, as in so many other cases, strikingly displayed. Tycho was not aware of the inroads which both his labours and his disappointments had made upon his constitution. Though surrounded with affectionate friends and admiring disciples, he was still an exile in a foreign land. Though his country had been base in its ingratitude, it was yet the land which he loved,—the scene of his earliest affections, —the theatre of his scientific glory. These feelings continually preyed upon his mind, and his unsettled spirit was ever hovering among his native mountains. In this condition he was attacked with a disease of the most painful kind; and though the paroxysms of his agonies had lengthened intermissions, yet he saw that death was approaching. He implored his pupils to persevere in their scientific labours. He conversed with Kepler on some of the profoundest points of astronomy, and with these secular occupations he mingled frequent acts of piety and devotion. In this happy condition he expired without pain at the age of fifty-five, unquestionably the victim of the councillors of Christian IV.

Notwithstanding the accessions which astronomy had received from the labours of Copernicus and Tycho, yet no progress was yet made in developing the general laws of the system, and scarcely an idea had been formed of the power by which the planets were retained in their orbits. The labours of assiduous observers had supplied the materials for this purpose, and Kepler arose to lay the foundations of physical astronomy.

John Kepler was born at Wiel, in Wirtemberg, in 1571. He was educated for the church, and even discharged some of the clerical functions; but his devotion to science withdrew him from the study of theology. Having received mathematical instruction from the celebrated Mæstlin, he had made such progress in the science, that he was invited in 1594 to fill the mathematical chair of Gratz in Styria. Endowed with a fertile imagination, his mind was ever intent upon subtle and ingenious speculations. In the year 1596, he published his peculiar views in a work on the Harmonies and Analogies of Nature. In this singular production, he attempts to solve what he calls the great cosmographical mystery of the admirable proportion of the planetary orbits; and by means of the six regular geometrical solids,* he endeavours to assign a reason why there are six planets, and why the dimensions of their orbits, and the time of their periodical revolutions, were such as Copernicus had found them. If a cube, for example, were inserted in a sphere, of which Saturn's orbit was one of the great circles, it would, he supposed, touch by its six planes the lesser sphere of Jupiter; and, in like manner, he proposes to determine, by the aid of the other geometrical solids, the magnitude

---

* The cube, the sphere, the tetrahedron, the octohedron, the dodecahedron, and the icosahedron.

of the spheres of the other planets. A copy of this work was presented by its author to Tycho Brahe, who had been too long versed in the severe realities of observation, to attach any value to such wild theories. He advised his young friend, "first to lay a solid foundation for his views by actual observation, and then, by ascending from these, to strive to reach the causes of things;" and there is reason to think, that, by the aid of the whole Baconian philosophy, thus compressed by anticipation into a single sentence, he abandoned for a while his visionary inquiries.

In the year 1598, Kepler suffered persecution for his religious principles, and was compelled to quit Gratz; but though he was recalled by the States of Styria, he felt his situation insecure, and accepted of a pressing invitation from Tycho to settle at Prague, and assist him in his calculations. Having arrived in Bohemia in 1600, he was introduced by his friends to the Emperor Rodolph, from whom he ever afterwards received the kindest attention. On the death of Tycho in 1601, he was appointed mathematician to the emperor, a situation in which he was continued during the successive reigns of Matthias and Ferdinand; but what was of more importance to science, he was put in possession of the valuable collection of Tycho's observations. These observations were remarkably numerous; and as the orbit of Mars was more oval than that of any of the other planets, they were peculiarly suitable for determining its real form. The notions of harmony and symmetry in the construction of the solar system, which had filled the mind of Kepler, necessarily led him to believe that the planets revolved with an uniform motion in circular orbits. So firm, indeed, was this conviction, that he made numerous attempts to represent the observations of Tycho by this hypothesis. The deviations were too great to be ascribed to errors of observation; and in trying various other curves, he was led to the discovery, that

Mars revolved round the sun in an elliptical orbit, in one of the foci of which the sun itself was placed. The same observations enabled him to determine the dimensions of the planet's orbit, and by comparing together the times in which Mars passed over different portions of its orbit, he found that they were to one another, as the areas described by the lines drawn from the centre of the planet to the centre of the sun; or, in more technical terms, that the radius vector describes equal areas in equal times. These two remarkable discoveries, the first that were ever made in physical astronomy, were extended to all the other planets of the system, and were communicated to the world in 1609, in his "Commentaries on the Motions of the Planet Mars, as deduced from the Observations of Tycho Brahe."

Although our author was conducted to these great laws by the patient examination of well-established facts, his imagination was ever hurrying him among the wilds of conjecture. Convinced that the mean distances of the planets from the sun bore to one another some mysterious relation, he not only compared them with the regular geometrical solids, but also with the intervals of musical tones; an idea which the ancient Pythagoreans had suggested, and which had been adopted by Archimedes himself. All these comparisons were fruitless; and Kepler was about to abandon an inquiry of about seventeen years' duration, when, on the 8th of March, 1618, he conceived the idea of comparing the powers of the different numbers which express the planetary distances, in place of the numbers themselves. He compared the squares and the cubes of the distances with the same powers of the periodic times; nay, he tried even the squares of the times with the cubes of the distances; but his hurry and impatience led him into an error of calculation, and he rejected this law as having no existence in

nature! On the 15th of May, his mind again reverted
to the same notion, and upon making the calculations
anew, and free from error, he discovered the great law,
that the squares of the periodic times of any two planets
are to one another as the cubes of their distances from
the sun. Enchanted with this unexpected result, he could
scarcely trust his calculations ; and, to use his own
language, he at first believed that he was dreaming, and
had taken for granted the very truth of which he was in
search. This brilliant discovery was published in 1619,
in his "Harmony of the World," a work dedicated to
James VI. of Scotland. Thus were established what have
been called the three laws of Kepler,—the motion of the
planets in elliptical orbits; the proportionality between the
areas described and their times of description; and the
relations between the squares of the periodic times and
the cubes of the distances.

The relation of the movements of the planets to the sun,
as the general centre of all their orbits, could not fail to
suggest to Kepler that some power resided in that luminary
by which these various motions were produced ; and he
went so far as to conjecture, that this power diminishes
as the square of the distance of the body on which it is
exerted ; but he immediately rejected this law, preferring
that of the simple distances. In his work on Mars, he
speaks of gravity as a mutual and corporeal affection
between similar bodies. He maintained also, that the tides
were occasioned by the moon's attraction, and that the
irregularities of the lunar motions, as detected by Tycho,
were owing to the joint actions of the sun and the earth.
Our countryman, Dr. Gilbert, in his celebrated book,
De Magnete, published in 1600, had about the same time
announced similar opinions on gravitation. He compared
the earth's action on the moon to that of a great loadstone,
and in his posthumous work which appeared half a century

afterwards, he maintains that the earth and moon act upon each other like two magnets, the influence of the earth being the greater on account of its superior mass. But though these opinions were a step in celestial physics, yet the identity of the gravity which is exhibited on the earth's surface by falling bodies with that which guides the planets in their orbits was not revealed either to the English or the German philosopher. It required more patience of thought than either could command, and its discovery was reserved for the exercise of higher powers.

The misery in which Kepler lived forms a painful contrast with the services which he performed to science. The pension on which he subsisted was always in arrears; and though the three emperors, whose reigns he adorned, directed their ministers to be more punctual in its payment, the disobedience of their commands was a source of continued vexation to Kepler. When he retired to Sagan, in Silesia, to spend in retirement the remainder of his days, his pecuniary difficulties became still more harassing. Necessity at last compelled him to apply personally for the arrears which were due; and he accordingly set out in 1630 for Ratisbon; but in consequence of the great fatigue which so long a journey on horseback produced, he was seized with a fever, which carried him off on the 30th of November, 1630, in the fifty-ninth year of his age. Thus perished one of the noblest of his race, a victim of poverty, and a martyr to science.

While Kepler was thus laying the foundation of physical astronomy, Galileo was busily employed in extending the boundaries of the solar system. This distinguished philosopher was born at Pisa in 1564. He was the son of a Florentine nobleman, and was educated for the medical profession; but a passion for geometry took possession of his mind, and called forth all his powers. Without the aid of a master, he studied the writings of Euclid and of

Archimedes; and such were his acquirements, that he was appointed by the Grand Duke of Tuscany to the mathematical chair of Pisa in the twenty-fifth year of his age. His opposition to the Aristotelian philosophy gained him many enemies, and at the end of three years he quitted Pisa, and accepted of an invitation to the professorship of mathematics at Padua. Here he continued for eighteen years adorning the university by his name, and diffusing around him a taste for the physical sciences. With the exception of some contrivances of inferior importance, Galileo had distinguished himself by no discovery till he had reached the forty-fifth year of his age. In the year 1609, the memorable year in which Kepler published his "Astronomia Nova," Galileo paid a visit to Venice, where he heard in the course of conversation, that a Dutchman of the name of Jansens had constructed and presented to Prince Maurice an instrument through which he saw distant objects magnified and rendered more distinct, as if they had been brought nearer to the observer. This report was credited by some and disbelieved by others; but, in the course of a few days, Galileo received a letter from James Badovere at Paris, which placed beyond a doubt the existence of such an instrument. The idea instantly filled his mind as one of the utmost importance to science; and so thoroughly was he acquainted with the properties of lenses, that he not only discovered the principle of its construction, but was able to complete a telescope for his own use. Into one end of a leaden tube he fitted a spectacle-glass, plane on one side and convex on the other; and in the other end he placed another spectacle-glass, concave on one side and plane on the other. He then applied his eye to the concave glass, and saw objects "pretty large and pretty near him." They appeared three times nearer, and nine times larger in surface, than to the naked eye. He soon after made another, which represented objects above sixty times larger; and, sparing neither labour nor expense, he finally constructed an instrument so excellent, as "to show things almost a thousand times larger (in surface), and above thirty times nearer to the naked eye."

There is, perhaps, no invention that science has presented to man so extraordinary in its nature, and so boundless in its influence, as that of the telescope. To the uninstructed mind, the power of seeing an object a thousand miles distant, as large and nearly as distinct as if it were brought within a mile of the observer, must seem almost miraculous; and even to the philosopher, who thoroughly comprehends the principles upon which it acts, it must ever appear one of the most elegant applications of science. To have been the first astronomer in whose hands such a power was placed, was a preference to which Galileo has owed much of his reputation.

Before the telescope was directed to the heavens, it was impossible to distinguish a planet from a star. But even with his first instrument, Galileo saw that Jupiter had a round appearance like the sun and moon; and on the 7th of January, 1610, when he used a telescope of superior power, he saw three little bright stars very near the planet, two on the east and one on the west side of it, all three situated in a line parallel to the ecliptic. Regarding them as ordinary stars, he never thought of estimating their distances. On the following day, when he accidentally directed his telescope to Jupiter, he was surprised to see the three stars to the west of the planet. To produce this effect it was requisite that the motion of Jupiter should be direct, though, according to calculation, it was actually retrograde. In this dilemma he waited with patience for the evening of the 9th, but unfortunately the sky was covered with clouds. On the 10th he saw only two stars to the east,—a circumstance which he was no longer able

to explain by the motion of Jupiter. He was therefore
compelled to ascribe the change to the stars themselves;
and, upon repeating his observations on the 11th, he no
longer doubted that he had discovered three secondary
planets revolving round Jupiter. On the 13th of January,
he for the first time saw the fourth satellite.

This discovery, of the utmost importance in itself,
derived an additional value from the light which it threw
on the true system of the universe. While the earth
was supposed to be the only planet enlightened by a moon,
it might naturally be regarded as alone habitable, and
therefore entitled to the pre-eminence of occupying the
centre of the system; but the discovery of four moons
round a much larger planet deprived this argument of its
force, and created a new analogy between the earth and
the other planets. When Kepler received the " Sidereal
Messenger," the work in which Galileo announced his
discovery in 1610, he perused it with the deepest interest;
and while it confirmed and extended his substantial discoveries, it dispelled at the same time some of those har-
monic dreams which still hovered among his thoughts.
In the "Dissertation" which he published on the discovery
of Galileo, he expresses his hope that satellites will be
discovered round Saturn and Mars—he conjectures that
Jupiter has a motion of rotation about his axis,—and
states his surprise, that, after what had been written on
the subject of telescopes by Baptista Porta, they had not
been earlier introduced into observatories.

In continuing his observations, Galileo applied his telescope to Venus, and in 1610 he discovered the phases of
that planet, which exhibited to him the various forms of
the waxing and the waning moon. This fact established
beyond a doubt that the planet revolved round the sun,
and thus gave an additional blow to the Ptolemaic system.
In his observations on the sun, Galileo discovered his spots,

and deduced from them the rotation of the central luminary. He observed that the body of Saturn had "handles" attached to it; but he was unable to detect the form of its ring, or render visible its minute satellites. On the surface of the moon he discovered her mountains and valleys, and determined the curious fact of her libration, in virtue of which parts of the margin of her disc occasionally appear and disappear. In the Milky Way he descried numerous minute stars which the unassisted eye was unable to perceive; and as the largest fixed stars, in place of being magnified by the telescope, became actually minute brilliant points, he inferred their immense distance as rendered necessary by the Copernican hypothesis. All his discoveries, indeed, furnished fresh arguments in favour of the new system; and the order of the planets, and their relation to a central sun, may now be considered as established by incontrovertible evidence.

While Galileo was occupied with these noble pursuits at Pisa, to which he had been recalled in 1611, his generous patron, Cosmo II., Grand Duke of Tuscany, invited him to Florence, that he might pursue with uninterrupted leisure his astronomical observations, and carry on his correspondence with the German astronomers. His fame had now resounded through all Europe;—the strongholds of prejudice and ignorance were unbarred;—and the most obstinate adherents of ancient systems acknowledged the meridian power of the day-star of science. Galileo was ambitious of propagating the great truths which he contributed so powerfully to establish. He never doubted that they would be received with gratitude by all,—by the philosopher as the consummation of the greatest efforts of human genius, and by the Christian as the most transcendent displays of Almighty Power. But he had mistaken the disposition of his species, and the character of the age. That same system of the heavens which had been discovered by

the humble ecclesiastic of Frauenberg, which had been patronized by the kindness of a bishop, and published at the
expense of a cardinal, and which the Pope himself had
sanctioned by the warmest reception, was, after the lapse
of a hundred years, doomed to the most violent opposition,
as subversive of the doctrines of the Christian faith.   On
no former occasion has the human mind exhibited such a
fatal relapse into intolerance.   The age itself had improved
in liberality;—the persecuted doctrines themselves had
become more deserving of reception;—the light of the
reformed faith had driven the Catholics from some of
their most obnoxious positions;—and yet, under all these
circumstances, the Church of Rome unfurled her banner
of persecution against the pride of Italy, against the ornament of his species, and against truths immutable and
eternal.

In consequence of complaints laid before the Holy
Inquisition, Galileo was summoned to appear at Rome in
1615, to answer for the heretical opinions which he had
promulgated.   He was charged with "maintaining as true
the false doctrine held by many, that the sun was immovable in the centre of the world, and that the earth
revolved with a diurnal motion;—with having certain
disciples to whom he taught the same doctrine;—with
keeping up a correspondence on the subject with several
German mathematicians;—with having published letters
on the solar spots, in which he explained the same doctrine
as true;—and with having glossed over with a false interpretation the passages of Scripture which were urged
against it."   The consideration of these charges came
before a meeting of the Inquisition, which assembled on
the 25th of February, 1616; and the court, declaring their
disposition to deal gently with the prisoner, pronounced
the following decree:—"That Cardinal Bellarmine should
enjoin Galileo to renounce entirely the above-recited false

opinions; that, on his refusal to do so, he should be commanded by the commissary of the Inquisition to abandon the said doctrine, and to cease to teach and defend it; and that, if he did not obey this command, he should be thrown into prison." On the 26th of February, Galileo appeared before Cardinal Bellarmine, and, after receiving from him a gentle admonition, he was commanded by the commissary, in the presence of a notary and witnesses, to desist altogether from his erroneous opinions; and it was declared to be unlawful for him in future to teach them in any way whatever, either orally or in his writings. To these commands Galileo promised obedience, and was dismissed from the Inquisition.

The mildness of this sentence was no doubt partly owing to the influence of the Grand Duke of Tuscany, and other persons of rank and influence at the Papal Court, who took a deep interest in the issue of the trial. Dreading, however, that so slight a punishment might not have the effect of putting down the obnoxious doctrines, the Inquisition issued a decree denouncing the new opinions as false and contrary to the sacred writings, and prohibiting the sale of every book in which they should be maintained.

Thus liberated from his persecutors, Galileo returned to Florence, where he pursued his studies with his wonted diligence and ardour. The recantation of his astronomical opinions was so formal and unreserved, that ordinary prudence, if not a sense of personal honour, should have restrained him from unnecessarily bringing them before the world. No anathema was pronounced against his scientific discoveries; no interdict was laid upon the free exercise of his genius. He was prohibited merely from teaching a doctrine which the Church of Rome considered to be injurious to its faith. We might have expected, therefore, that a philosopher so conspicuous in the eyes of the world would have respected the prejudices, however

base, of an institution whose decrees formed part of the
law of the land, and which possessed the power of life and
death within the limits of its jurisdiction.   Galileo, however, thought otherwise.   A sense of degradation* seems
to have urged him to retaliate, and before six years had
elapsed he began to compose his " Cosmical System, or
Dialogues on the two greatest Systems of the World, the
Ptolemæan and the Copernican," the concealed object of
which is to establish the opinions which he had promised
to abandon.   In this work the subject is discussed by three
speakers, Sagredo, Salviatus, and Simplicius, a peripatetic
philosopher, who defends the system of Ptolemy with
much skill against the overwhelming arguments of the
rival disputants.   Galileo hoped to escape notice by this
indirect mode of propagating the new system, and he
obtained permission to publish his work, which appeared
at Florence in 1632.

The Inquisition did not, as might have been expected,
immediately summon Galileo to their presence.   Nearly a
year elapsed before they gave any indication of their
design ; and, according to their own statement, they did
not even take the subject under consideration till they saw
that the obnoxious tenets were every day gaining ground,
in consequence of the publication of the Dialogues.   They
then submitted the work to a careful examination, and,
having found it to be a direct violation of the injunction
which had been formerly intimated to its author, they

* It is distinctly stated in the sentence of the Inquisition, that Galileo's
enemies had charged him with having abjured his opinions in 1616, and
affirmed that he had been punished by the Inquisition. In order to refute
these calumnies, Galileo applied to Cardinal Bellarmine for a certificate to
prove that he neither abjured his opinions, nor suffered any punishment
for them; but that the doctrine of the motion of the earth and the
stability of the sun was only denounced to him as contrary to Scripture,
and as one which could not be defended or maintained. Cardinal Bellarmine drew up such a certificate in his own handwriting.

again cited him before their tribunal in 1633. The venerable sage, now in his seventieth year, was thus compelled
to repair to Rome, and when he arrived he was committed
to the apartments of the Fiscal of the Inquisition. The
unchangeable friendship, however, of the Grand Duke of
Tuscany obtained a remission of this severity, and Galileo
was allowed to reside at the house of the Tuscan ambas-
ador during the two months which the trial occupied.
When brought before the Inquisition, and examined upon
oath, he acknowledged that the Dialogues were written
by himself, and that he obtained permission to publish
them without notifying to the person who gave it, that he
had been prohibited from holding, defending, or teaching
the heretical opinions. He confessed also, that the Dialogues were composed in such a manner, that the arguments in favour of the Copernican system, though given
as partly false, were yet managed in such a manner, that
they were more likely to confirm than overturn its doctrines; but stated that this error, which was not inten-
tional, arose from the natural desire of making an
ingenious defence of false propositions, and of opinions·
that had the semblance of probability.

After receiving these confessions and excuses, the
Inquisition allowed Galileo a proper time for giving in his
defence; but this seems to have consisted solely in bringing forward the certificate of Cardinal Bellarmine, already
mentioned, which made no allusion to the promise under
which Galileo had placed himself never to defend, or
teach in any way whatever, the Copernican doctrines.
The court held this defence to be an aggravation of the
crime rather than an excuse for it, and proceeded to pronounce a sentence which will ever be memorable in the
history of the human mind.

Invoking the name of our Saviour, they declare, that
Galileo had rendered himself liable to the suspicion of

heresy, by believing the doctrine, contrary to Scripture, that the sun was the centre of the earth's orbit, and did not move from east to west; and by defending as probable the opinion, that the earth moved, and was not the centre of the world; and that he had thus incurred all the censures and penalties which were enacted by the church against such offences;—but that he should be absolved from these penalties, provided he sincerely abjured and cursed all the errors and heresies alluded to in the formula of the church, which should be submitted to him. That so grave and pernicious a crime should not pass altogether unpunished, —that he might become more cautious in future,—and might be an example to others to abstain from such offences, they decreed that his Dialogues should be prohibited by a formal edict,—that he should be condemned to the prison of the Inquisition during pleasure,—and that, during the three following years, he should recite once a week the seven penitential psalms.

This sentence was subscribed by seven cardinals; and on the 22nd of June, 1633, Galileo signed an abjuration, humiliating to himself and degrading to philosophy. At the age of seventy, on his bended knees, and with his right hand resting on the Holy Evangelists, did this patriarch of science avow his present and his past belief in all the dogmas of the Romish Church, abandon as false and heretical the doctrine of the earth's motion, and of the sun's immobility, and pledge himself to denounce to the Inquisition any other person who was even suspected of heresy. He abjured, cursed, and detested those eternal and immutable truths which the Almighty had permitted him to be the first to establish. What a mortifying picture does this scene present to us of moral infirmity and intellectual weakness! If the unholy zeal of the assembly of cardinals has been justly branded with infamy, what must we think of the venerable sage whose grey hairs were entwined with the chaplet of immortality, quailing under the fear of man, and sacrificing the convictions of his conscience, and the deductions of his reason, at the altar of a base superstition? Had Galileo added the courage of the martyr to the wisdom of the sage;—had he carried the glance of his indignant eye round the circle of his judges, and, with uplifted hands, called upon the living God to witness the truth and immutability of his opinions, the bigotry of his enemies would have been disarmed, and science would have achieved a memorable triumph.

The great truth of the Copernican system, instead of being considered as heretical, had been actually adopted by many pious members of the Catholic Church, and even some of its dignitaries did not scruple to defend it openly. Previous to the first persecution of Galileo in 1615, a Neapolitan nobleman, Vincenzio Caraffa, a person equally distinguished by his piety and birth, had solicited Paul Anthony Foscarinus, a learned Carmelite monk, to illustrate and defend the new system of the universe. With this request the ecclesiastic speedily complied; and in the pamphlet, which he completed on the 6th of January, 1615, he defends the Copernican system with much boldness and ingenuity; he reconciles the various passages of Scripture with the new doctrine, and he expresses the hope that such an attempt, now made for the first time, will prove agreeable to philosophers, but particularly to those very learned men, Galileo Galilei, John Kepler, and all the members of the Lyncean Academy, who, he believes, entertain the same opinion. This remarkable production, written from the Convent of the Carmelites at Naples, is dedicated to the very Reverend Sebastian Fantoni, general of the order of Carmelites, and was published at Florence, with the sanction of the ecclesiastical authorities, in 1630; three years before the second persecution of Galileo.

It would be interesting to know the state of public feeling in Italy when Galileo was doomed to the prisons of the Inquisition. No appeal seems to have been made against so cruel a sentence; and neither in remonstrance nor in derision does an individual voice seem to have been raised. The master spirits of the age looked with sullen indifference on the persecution of exalted genius; and Galileo lay in chains deserted and unpitied. This unrebuked triumph of his enemies was perhaps favourable to the object of their vengeance. Resistance might have heightened the rigour of a sentence which submission seems to have alleviated. The interference of some eminent individuals of Rome, among whom we have no doubt that the Grand Duke of Tuscany was the most influential, induced Pope Urban VIII. not only to shorten the period, but to soften the rigour, of Galileo's imprisonment. From the dungeon of the Inquisition, where he had remained only four days, he was transported to the ambassador's palace in the Garden de Medici at Rome; and when his health had begun to suffer, he was permitted to leave the metropolis; and would have been allowed to return to Florence, but as the plague raged in that city, he was sent, in July, 1633, to the archiepiscopal palace of Sienna, the residence of the Archbishop Piccolomini, where he carried on and completed his valuable investigations respecting the resistance of solids. Here he continued five months, when, in consequence of the disappearance of the plague at Florence, he was allowed to retire to his villa at Bellosguardo, and afterwards to that of Arcetri* in the vicinity of Florence.

* The place where now stands the magnificent new Royal Observatory, inaugurated with much ceremony in the autumn of 1872, and placed under the directorship of Professor Donati, who unfortunately died in September of the following year. He was shortly afterwards succeeded at the Observatory by Schiaparelli (the present Director), who had already acquired for himself a distinguished position in astronomy by his labours and discoveries at Milan.—EDITOR.

Though Galileo was now, to a certain degree, liberated from the power of man, yet the afflicting dispensations of Providence began to fall thickly around him. No sooner had he returned to Arcetri, than his favourite daughter, Maria, was seized with a dangerous illness, which soon terminated in her death. He was himself attacked with hernia, palpitation of the heart, loss of appetite, and the most oppressive melancholy; and though he solicited permission to repair to Florence for medical assistance, yet this small request was refused him. In 1638, however, the Pope permitted him to pay a visit to Florence; and his friend, Father Castelli, was allowed to visit him in the company of an officer of the Inquisition. But this indulgence was soon withdrawn, and at the end of a few months he was remanded to Arcetri. The sight of his right eye had begun to fail in 1636, from an opacity of the cornea. In 1637 his left eye was attacked with the same complaint; so that in a few months he was affected with total and incurable blindness. Before this calamity had supervened, he had noticed the curious phenomenon of the moon's libration, in consequence of which, parts of her visible disc that are exposed to view at one time are withdrawn at another. He succeeded in explaining two of the causes of this curious phenomenon, viz., the different distances of the observer from the line joining the centre of the earth and the moon, which produces the diurnal libration; and the unequal motion of the moon in her orbit, which produces the libration in longitude. It was left, however, to Hevelius to discover the libration in latitude, which arises from the inclination of her axis being a little less than a right angle to the ecliptic; and to Lagrange to discover the spheroidal libration, or that which arises from the action of the earth upon the lunar spheroid.

The sorrows with which Galileo was now beset, seem to have disarmed the severity of the Inquisition. He was freely permitted to enjoy the society of his friends, who now thronged around him to express their respect and their sympathy. The Grand Duke of Tuscany was his frequent visitor, and Gassendi, Diodati, and our countryman Milton, went to Italy for the purpose of visiting him. He entertained his friends with the warmest hospitality; and though simple and abstemious in his diet, yet he was fond of good wine, and seems even in his last days to have paid particular attention to the excellence of his cellar.

Although Galileo had nearly lost his hearing as well as his sight, yet his intellectual faculties were unimpaired; and while his mind was occupied in considering the force of percussion, he was seized with fever and palpitation of the heart, which, after two months' illness, terminated his life on the 8th of January, 1642.

Among the predecessors of Newton in astronomical research, we must not omit the names of Bouillaud, (Bullialdus,) Borelli, and Dr. Hooke. Ismael Bouillaud, a native of Laon in France, and the author of several valuable astronomical works, has derived more reputation from a single sentence in his *Astronomica Philolaica*, published in 1645, than from all the rest of his labours. He was not a believer in the doctrine of attraction, which, as we have already seen, had been broached by Copernicus, and discovered by Kepler; but in speaking of that power as the cause of the planetary motions, he remarks, "that if attraction existed, it would decrease as the square of the distance." The influence of gravity was still more distinctly developed by Borelli,* an Italian philosopher, who published at Florence, in 1666, a work on Jupiter's satellites, in which he maintains, that all the planets perform their motions round the sun according to a general law; that the satellites of Jupiter and of Saturn move round their

---

* Born at Naples in 1608; died at Rome in 1679. Was Professor of Mathematics at Messina, and afterwards at Pisa.—EDITOR.

primary planets in the same manner as the moon does round the earth ; and that they all revolve round the sun, which is the only source of any virtue, and that this virtue attaches them and unites them so that they cannot recede from their centre of action.*

Our countryman, Dr. Robert Hooke, seems to have devoted much of his attention to the cause of the planetary motions. On the 21st of March, 1666, he read to the Royal Society an account of a series of experiments for determining whether bodies experience any variation in their weight at different distances from the centre of the earth. His experiments, as Hooke himself saw, were by no means satisfactory, and hence he was led to the ingenious idea of measuring the force of gravity by observing, at different altitudes, the rate of a pendulum clock. About two months afterwards, he exhibited to the Society an approximate representation of the forces which retain the planets in their orbits, in the paths described by a circular pendulum impelled with different degrees of force; but though this experiment illustrated the production of a curvilineal motion, by combining a tangential force with a central power of attraction, yet it was only an illustration, and could not lead to the true cause of the planetary motions. At a later period, however, viz., in 1674, Hooke resumed the subject in a dissertation entitled, "An attempt to prove the Motion of the Earth from Observation," which

* M. Delambre maintains that these views of Borelli are only those of Kepler, slightly modified. Newton and Huygens have attached to them a greater value. The last of these philosophers remarks, "Refert Plutarchus, fuisse jam olim qui putaret ideo manere lunam in orbe suo, quòd vis recedendi a terrâ, ob motum circularem, inhiberetur pari vi gravitatis quâ ad terram accedere conaretur. Idemque ævo nostro, non de lunâ tantum sed et planetis cæteris statuit Alphonsus Borellus, ut nempe primariis eorum gravitas esset solem versus, hunis vero ad terram, Jovem ac Saturnum quos comitantur."—Huygens, Cosmothoor., lib. ii., Opera, t. ii., p. 720.

contains the following remarkable observations upon gravity.

"I shall hereafter explain a system of the world, differing in many particulars from any yet known, answering in all things to the common rules of mechanical motions. This depends upon three suppositions:—first, that all celestial bodies whatsoever have an attraction or gravitating power towards their own centres, whereby they attract, not only their own parts, and keep them from flying from them, as we may observe the earth to do, but that they also do attract all the other celestial bodies that are within the sphere of their activity, and consequently, that not only the sun and moon have an influence upon the body and motion of the earth, and the earth upon them, but that Mercury, Venus, Mars, Jupiter, and Saturn also, by their attractive powers, have a considerable influence upon every one of their motions also. The second supposition is this, that all bodies whatsoever, that are put into a direct and simple motion, will so continue to move forward in a straight line, till they are, by some other effectual powers, deflected, and sent into a motion describing a circle, ellipsis, or some other more compounded curve line. The third supposition is, that those attractive powers are so much the more powerful in operating by how much the nearer the body wrought upon is to their own centres. *Now, what these several degrees are I have not yet experimentally verified;* but it is a notion which, if fully prosecuted, as it ought to be, will mightily assist the astronomers to reduce all the celestial motions to a certain rule, which I doubt will never be done without it. He that understands the nature of the circular pendulum and circular motion, will easily understand the whole of this principle, and will know where to find directions in nature for the true stating thereof. This I only hint at present to such as have ability and opportunity of prosecuting this inquiry,

and are not wanting of industry for observing and calculating, wishing heartily such may be found, having myself
many other things in hand, which I would first complete,
and therefore cannot so well attend it.   But this I do
promise the undertaker, that he will find all the great
motions of the world to be influenced by this principle,
and that the true understanding thereof will be the true
perfection of astronomy."

In this remarkable passage, the doctrine of universal
gravitation and the general law of the planetary motions
are clearly laid down.   Delambre, we think, scarcely does
justice to Hooke, when he says that what it contains " is
to be found expressly in Kepler."*   Kepler did indeed
mention as probable the law of the squares of the distances,
but he afterwards, as Delambre admits, rejected it for that
of the simple distances. Hooke, on the contrary, announced
it as a truth; and in a letter which he addressed to Newton
in 1679, relative to the curve described by a projectile influenced by the earth's daily motion, he asserted that if the
force of gravity decreased as the square of the distance,
the curve described by a projectile would be an ellipse,
whose focus was the centre of the earth.   But however
great be the merit which we may assign to Hooke's experimental results and sagacious views, they cannot be regarded either as anticipating the discoveries of Newton,
or diminishing his fame.   Newton had made the same
discoveries by independent researches; and there is no
reason to believe that he derived any of his ideas from his
contemporaries.

* *Histoire de l'Astronomie du dix-huitième Siècle*, p. 9.

CHAPTER XL.

The first Idea of Gravity occurs to Newton in 1666—His first Speculations upon it—Interrupted by his Optical Experiments—He resumes the Subject in consequence of a Discussion with Dr. Hooke—He discovers the true Law of Gravity and the Cause of the Planetary Motions—Dr. Halley urges him to publish his Principia—His Principles of Natural Philosophy—Proceedings of the Royal Society on this Subject—The Principia appears in 1687—General Account of it, and of the Discoveries it contains—They meet with great Opposition, owing to the Prevalence of the Cartesian System—Account of the Reception and Progress of the Newtonian Philosophy in Foreign Countries—Account of its Progress and Establishment in England.

SUCH is a brief sketch of the labours and lives of those illustrious men who prepared the science of astronomy for the application of Newton's genius. Copernicus had determined the arrangement and general movements of the planetary bodies: Kepler had proved* that they moved in elliptical orbits; that their *radii vectores* described arcs proportional to the times; and that the squares of their periodic times are as the cubes of their mean distances from the sun. Galileo had added to the universe a whole system of secondary planets; and several astronomers had

---

* Strictly speaking, Kepler had shown (and great had been the thought and labour employed in doing so) that these laws were true of the motions of Mars, and that the third resulted from a comparison of the orbits of the Earth and Mars. But by analogy "Kepler's laws" were extended to the other planets.—EDITOR.

distinctly referred the movements of the heavenly bodies to the power of attractive force producing curvilineal motion from motion in a straight line.

In the year 1666, when the plague had driven Newton from Cambridge, he was sitting alone in the garden at Woolsthorpe, and reflecting on the nature of gravity, that remarkable power which causes all bodies to descend towards the centre of the earth. As this power is not found to suffer any sensible diminution at the greatest distance from the earth's centre to which we can reach, being as powerful at the tops of the highest mountains as at the bottom of the deepest mines, he conceived it highly probable that it must extend much farther than was usually supposed. No sooner had this happy conjecture occurred to his mind, than he considered what would be the effect of its extending as far as the moon. That her motion must be influenced by such a power, he did not for a moment doubt; and a little reflection convinced him that it might be sufficient for retaining that luminary in her orbit round the earth. Though the force of gravity suffers no sensible diminution at those small distances from the earth's centre at which we can place ourselves, yet he thought it very possible, that, at the distance of the moon, it might differ much in strength from what it is on the earth. In order to form some estimate of the degree of its diminution, he considered, that, if the moon be retained in her orbit by the force of gravity, the primary planets must also be carried round the sun by the same power; and by comparing the periods of the different planets with their distances from the sun, he found, that, if they were retained in their orbits by any power like gravity, its force must decrease in the duplicate proportion,* or as the

* "But for the duplicate proportion, I gathered it from Kepler's theorem about twenty years ago."—Newton's Letter to Halley, July 14th, 1686.

squares of their distances from the sun. In drawing this
conclusion, he supposed the planets to move in orbits perfectly circular, and having the sun in their centre. Having
thus obtained the law of the force by which the planets
were drawn to the sun, his next object was to ascertain if
such a force emanating from the earth, and directed to the
moon, was sufficient, when diminished in the duplicate
ratio of the distance, to retain her in her orbit. In performing this calculation, it was necessary to compare the
space through which heavy bodies fall in a second at a
given distance from the centre of the earth, viz., at its
surface, with the space through which the moon, as it
were, falls to the earth in a second of time, while revolving in a circular orbit. Being at a distance from books
when he made this computation, he adopted the common
estimate of the earth's diameter then in use among geo-
graphers and navigators, and supposed that each degree of
latitude contained 60 English miles. In this way he found
that the force which retains the moon in her orbit, as
deduced from the force which occasions the fall of heavy
bodies to the earth's surface, was *one-sixth* greater than
that which is actually observed in her circular orbit. This
difference threw a doubt upon all his speculations; but,
unwilling to abandon what seemed to be otherwise so
plausible, he endeavoured to account for the difference of
the two forces, by supposing that some other cause* must
have been united with the force of gravity in producing so
great a velocity of the moon in her circular orbit. As this
new cause, however, was beyond the reach of observation,
he discontinued all further inquiries into the subject, and
concealed from his friends the speculations in which he
had been employed.

* Whiston asserts that this cause was supposed by Newton to be
something analogous to the vortices of Descartes. — See Whiston's
Memoirs of Himself, p. 231.

After his return to Cambridge in 1666, his attention was occupied with those optical discoveries of which we have given an account in a preceding chapter; but he had no sooner brought them to a close than his mind reverted to the great subject of the planetary motions. Upon the death of Oldenburg in August, 1678, Dr. Hooke was appointed secretary to the Royal Society; and as this learned body had requested the opinion of Newton about a system of physical astronomy, he addressed a letter to Dr. Hooke on the 28th November, 1679. In this letter he proposed a direct experiment for verifying the motion of the earth, viz., by observing whether or not bodies that fall from a considerable height descend in a vertical direction; for if the earth was at rest the body would describe exactly a vertical line, whereas, if it revolved round its axis, the falling body must deviate from the vertical line towards the east. The Royal Society attached great value to the idea thus casually suggested, and Dr. Hooke was appointed to put it to the test of experiment. Being thus led to consider the subject more attentively, he wrote to Newton, that wherever the direction of gravity was oblique to the axis on which the earth revolved, that is, in every part of the earth except the equator, falling bodies should approach to the equator, and the deviation from the vertical, in place of being exactly to the east, as Newton maintained, should be to the south-east of the point from which the body began to move. Newton acknowledged that this conclusion was correct in theory, and Dr. Hooke is said to have given an experimental demonstration of it before the Royal Society in December, 1679.* Newton had erroneously concluded, that the path of the falling body would be a spiral; but Dr. Hooke, on the same occasion on which he made the preceding experiment, read a

* Waller's *Life of Hooke*, p. 22.

paper to the Society, in which he proved that the path of
the body would be an eccentric ellipse in vacuo, and an
ellipti-spiral, if the body moved in a resisting medium.[*]

This correction of Newton's error, and the discovery
that a projectile would move in an elliptical orbit when
under the influence of a force varying in the inverse ratio
of the square of the distance, led Newton, as he himself
informs us in his letter to Halley,[†] to discover " the theorem by which he afterwards examined the ellipsis," and to
demonstrate the celebrated proposition, that a planet acted
upon by an attractive force varying inversely as the
square of the distance, will describe an elliptical orbit in
one of whose foci the attractive force resides.

But though Newton had thus discovered the true cause
of all the celestial motions, he did not yet possess any
evidence that such a force actually resided in the sun and
planets. The failure of his former attempt to identify the
law of falling bodies at the earth's surface with that which
guided the moon in her orbit, threw a doubt over all his
speculations, and prevented him from giving any account
of them to the public.

An accident, however, of a very interesting nature induced him to resume his former inquiries, and enabled him
to bring them to a close. In June, 1682, when he was
attending a meeting of the Royal Society of London, the
measurement of a degree of the meridian, executed by M.
Picard in 1679, became the subject of conversation.
Newton took a memorandum of the result obtained by the
French astronomer; and, having deduced from it the
diameter of the earth, he immediately resumed his calculation of 1665, and began to repeat it with these new data.
In the progress of the calculation he saw that the result

---

[*] Waller's *Life of Hooke*, p. 22.

[†] July 27th, 1686, *Biog. Brit.*, p. 2662.

which he had formerly expected was likely to be produced,
and he was thrown into such a state of nervous irritability
that he was unable to carry on the calculation. In this
state of mind he intrusted it to one of his friends, and he
had the high satisfaction of finding his former views amply
realized. The force of gravity which regulated the fall
of bodies at the earth's surface, when diminished as the
square of the moon's distance from the earth, was found
to be almost exactly equal to the centrifugal force of the
moon as deduced from her observed distance and velocity.

The influence of such a result upon such a mind may be
more easily conceived than described. The whole material
universe was spread out before him; the sun with all his
attending planets; the planets with all their satellites;
the comets wheeling in every direction in their eccentric
orbits; and the systems of the fixed stars stretching to the
remotest limits of space. All the varied and complicated
movements of the heavens, in short, must have been at
once presented to his mind as the necessary result of that
law which he had established in reference to the earth and
the moon.

After extending this law to the other bodies of the
system, he composed a series of propositions on the motion
of the primary planets about the sun, which were sent to
London about the end of 1683, and were soon afterwards
communicated to the Royal Society.*

About this period other philosophers had been occupied
with the same subject. Sir Christopher Wren had many
years before endeavoured to explain the planetary motions
" by the composition of a descent towards the sun, and an
impressed motion;" but he at length " gave it over, not
finding the means of doing it." In January, 1683—4, Dr.
Halley had concluded from Kepler's Law of the Periods

* *Commercium Epistolicum*, No. 71.

and Distances, that the centripetal force decreased in the reciprocal proportion of the squares of the distances; and having one day met Sir Christopher Wren and Dr. Hooke, the latter affirmed that he had demonstrated upon that principle all the laws of the celestial motions. Dr. Halley confessed that his attempts were unsuccessful, and Sir Christopher, in order to encourage the inquiry, offered to present a book of forty shillings' value to either of the two philosophers who should, in the space of two months, bring him a convincing demonstration of it. Hooke persisted in the declaration that he possessed the method, but avowed it to be his intention to conceal it for some time. He promised, however, to show it to Sir Christopher; but there is every reason to believe that this promise was never fulfilled.

In August, 1684, Dr. Halley went to Cambridge for the express purpose of consulting Newton on this interesting subject. Newton assured him that he had brought this demonstration to perfection, and promised him a copy of it. This copy was received in November by the Doctor, who made a second visit to Cambridge, in order to induce its author to have it inserted in the register book of the Society. On the 10th of December, Dr. Halley announced to the Society, that he had seen at Cambridge Mr. Newton's treatise *De Motu Corporum*, which he had promised to send to the Society to be entered upon their register; and Dr. Halley was desired to unite with Mr. Paget, Master of the Mathematical School in Christ's Hospital, in reminding Mr. Newton of his promise "for securing the invention to himself till such time as he can be at leisure to publish it." On the 25th of February, Mr. Aston, the Secretary, communicated a letter from Mr. Newton, in which he expressed his willingness "to enter in the register his notions about motion, and his intentions to fit them suddenly for the press." The progress

of his work was, however, interrupted by a visit of five or
six weeks which he made in Lincolnshire; but he proceeded with such diligence on his return, that he was able
to transmit the manuscript to London before the end of
April. This manuscript, entitled *Philosophiæ Naturalis
Principia Mathematica*, and dedicated to the Society, was
presented by Dr. Vincent on the 28th of April, 1686,
when Sir John Hoskins, the vice-president, and the particular friend of Dr. Hooke, was in the chair. Dr. Vincent
passed a just encomium on the novelty and dignity of the
subject; and another member added, that "Mr. Newton
had carried the thing so far, that there was no more to be
added." To these remarks the vice-president replied, that
the method "was so much the more to be prized as it was
both invented and perfected at the same time." Dr.
Hooke took offence at these remarks, and blamed Sir John
for not having mentioned "what he had discovered to
him;" but the vice-president did not seem to recollect
any such communication, and the consequence of this discussion was, that "these two, who till then were the most
inseparable cronies, have since scarcely seen one another,
and are utterly fallen out." After the breaking up of the
meeting, the Society adjourned to the coffee-house, where
Dr. Hooke stated that he not only had made the same
discovery, but had given the first hint of it to Newton.

An account of these proceedings was communicated to
Newton through two different channels. In a letter dated
May 22nd, Dr. Halley wrote to him " that Mr. Hooke had
some pretensions upon the invention of the rule of the
decrease of gravity being reciprocally as the squares of
the distances from the centre. He says you had the notion
from him, though he owns the demonstration of the curves
generated thereby to be wholly your own. How much of
this is so you know best, as likewise what you have to do
in this matter. Only Mr. Hooke seems to expect you will

make some mention of him in the preface, which it is possible you may see reason to prefix."

This communication from Dr. Halley induced our author, on the 20th of June, to address a long letter to him, in which he gives a minute and able refutation of Hooke's claims; but before this letter was dispatched, another correspondent, who had received his information from one of the members that were present, informed Newton "that Hooke made a great stir, pretending that he had all from him, and desiring they would see that he had justice done him." This fresh charge seems to have ruffled the tranquillity of Newton; and he accordingly added an angry and satirical postscript, in which he treats Hooke with little ceremony, and goes so far as to conjecture that Hooke might have acquired his knowledge of the law from a letter of his own to Huygens, directed to Oldenburg, and dated January 14th, 1672-3. "My letter to Hugenius was directed to Mr. Oldenburg, who used to keep the originals. His papers came into Mr. Hooke's possession. Mr. Hooke, knowing my hand, might have the curiosity to look into that letter, and there take the notion of comparing the forces of the planets arising from their circular motion; and so what he wrote to me afterwards about the rate of gravity, might be nothing but the fruit of my own garden."

In replying to this letter, Dr. Halley assured him that Hooke's "manner of claiming the discovery had been represented to him in worse colours than it ought, and that he neither made public application to the Society for justice, nor pretended that you had all from him." The effect of this assurance was to make Newton regret that he had written the angry postscript to his letter; and in replying to Halley on the 14th of July, 1686, he not only expresses his regret, but recounts the different new ideas which he had acquired from Hooke's correspondence, and suggests it

as the best method "of compromising the present dispute,"
to add a scholium, in which Wren, Hooke, and Halley are
acknowledged to have independently deduced the law of
gravity from the second law of Kepler.*

At the meeting of the 28th of April, at which the
manuscript of the *Principia* was presented to the Royal
Society, it was agreed that the printing of it should be
referred to the Council; that a letter of thanks should be
written to its author; and at a meeting of the Council on
the 19th of May, it was resolved that the MS. should be
printed at the Society's expense, and that Dr. Halley
should superintend it while going through the press.
These resolutions were communicated by Dr. Halley in a
letter dated the 22nd of May; and in Newton's reply
on the 20th of June, already mentioned, he makes the
following observations: "The proof you sent me I like
very well.   I designed the whole to consist of three books;
the second was finished last summer, being short, and only
wants transcribing, and drawing the cuts fairly.   Some
new propositions I have since thought on, which I can as
well let alone.   The third wants the theory of comets.
In autumn last I spent two months in calculation to no
purpose for want of a good method, which made me afterwards return to the first book, and enlarge it with diverse
propositions, some relating to comets, others to other things
found out last winter.   The third I now design to suppress.
Philosophy is such an impertinently litigious lady that a
man had as good be engaged in law-suits as have to do
with her.   I found it so formerly, and now I can no sooner
come near her again but she gives me warning.   The two
first books without the third, will not so well bear the
title of *Philosophiæ Naturalis Principia Mathematica;*
and therefore I had altered it to this, *De Motu Corporum*

* This Scholium is added to Prop. iv., lib. I., coroll. 6.

Libri duo. But, after second thoughts, I retain the former title. 'T will help the sale of the book, which I ought not to diminish now 't is yours."

In replying to this letter on the 29th of June, Dr. Halley regrets that our author's tranquillity should have been thus disturbed by envious rivals; and implores him in the name of the Society not to suppress the third book. "I must again beg you," says he, "not to let your resentments run so high as to deprive us of your third book, wherein your applications of your mathematical doctrine to the theory of comets, and several curious experiments which, as I guess by what you write, ought to compose it, will undoubtedly render it acceptable to those who will call themselves philosophers without mathematics, which are much the greater number."

To these solicitations Newton seems to have readily yielded. His second book was sent to the Society, and presented on the 2nd of March, 1686-7. The third book was also transmitted, and presented on the 6th of April, and the whole was completed and published in the month of May, 1687.

Such is a brief account of the publication of a work which is memorable not only in the annals of one science or of one country, but which will form an epoch in the history of the world, and will ever be regarded as the brightest page in the records of human reason. We shall endeavour to convey to the reader some idea of its contents, and of the brilliant discoveries which it disseminated over Europe.

The *Principia* consists of three books. The *first* and *second*, which occupy three-fourths of the work, are entitled, *On the Motion of Bodies*—the *first* treating of their motions in free space, and the *second* of their motions in a resisting medium; whilst the *third* bears the title, *On the System of the World.* The first two books contain the

mathematical Principles of Philosophy, namely, the laws
and conditions of motions and forces; and they are
illustrated with several philosophical scholia, which treat
of some of the most general and best established points
in philosophy, such as the density and resistance of bodies,
spaces void of matter, and the motion of sound and
light. The object of the third book is to deduce from
these principles the constitution of the system of the
world; and this book has been drawn up in as popular a
style as possible, in order that it may be generally read.

The great discovery which characterizes the *Principia* is
that of the principle of universal gravitation, as deduced
from the motion of the moon, and from the three great
facts or laws discovered by Kepler. This principle is,
*that every particle of matter in the universe is attracted by,
or gravitates to, every other particle of matter, with a force
inversely proportional to the squares of their distances.*
From the second law of Kepler, namely, the proportionality
of the areas to the times of their description, Newton
inferred that the force which keeps a planet in its orbit
is always directed to the sun; whilst from the first law
of Kep'er, that every planet moves in an ellipse with the
sun in one of its foci, he drew the still more general inference, that the force by which the planet moves round
that focus varies inversely as the square of its distance
from the focus. From the third law of Kepler, which
connects the distances and periods of the planets by a
general rule, Newton deduced the equality of gravity in
them all towards the sun, modified only by their different
distances from its centre; and in the case of terrestrial
bodies, he succeeded in verifying the equality of action by
numerous and accurate experiments.

By taking a more general view of the subject, Newton
demonstrated that a conic section was the only curve in
which a body could move when acted upon by a force

varying inversely as the square of the distance; and he established the conditions depending on the velocity and the primitive position of the body, which were requisite to make it describe a circular, an elliptical, a parabolic, or a hyperbolic orbit.

Notwithstanding the generality and importance of these results it still remained to be determined whether the force resided in the centres of the planets, or belonged to each individual particle of which they were composed. Newton removed this uncertainty by demonstrating that, if a spherical body acts upon a distant body with a force varying as the distance of this body from the centre of the sphere, the same effect will be produced as if each of its particles acted upon the distant body according to the same law. And hence it follows that the spheres, whether they are of uniform density or consist of concentric layers, with densities varying according to any law whatever, will act upon each other in the same manner as if their force resided in their centres alone. But as the bodies of the solar system are very nearly spherical, they will all act upon one another, and upon bodies placed on their surface, as if they were so many centres of attraction; and therefore we obtain the law of gravity which subsists between spherical bodies, namely, that one sphere will act upon another with a force directly proportional to the quantity of matter, and inversely as the square of the distance between the centres of the spheres. From the equality of action and reaction, to which no exception can be found, Newton concluded that the sun gravitates to the planets, and the planets to their satellites, and the earth itself to the stone which falls upon its surface; and consequently that the two mutually gravitating bodies approach one another with velocities inversely proportional to their quantities of matter.

Having established this universal law, Newton was enabled not only to determine the weight which the same body would have at the surface of the sun and the planets, but even to calculate the quantity of matter in the sun, and in all the planets that had satellites, and also to determine the density or specific gravity of the matter of which they were composed. In this way he found that the weight of the same body would be twenty-three times greater at the surface of the sun than at the surface of the earth, and that the density of the earth was four times greater than that of the sun; the planets increasing in density as they are nearer to the centre of the system.

If the peculiar genius of Newton has been displayed in his investigation of the law of universal gravitation, it shines with no less lustre in the patience and sagacity with which he traced the consequences of this fertile principle.

The discovery of the spheroidal form of Jupiter by Cassini had probably directed the attention of Newton to the determination of its cause, and consequently to the investigation of the true figure of the earth. The spherical form of the planets had been ascribed by Copernicus to the gravity or natural appetency of their parts; but upon considering the earth as a body revolving upon its axis, Newton quickly saw that the figure arising from the mutual attraction of its parts must be modified by another force arising from its rotation. When a body revolves upon an axis, the velocity of rotation increases from the poles where it is nothing, to the equator where it is a maximum. In consequence of this velocity the bodies on the earth's surface have a tendency to fly off from it, and this tendency increases with the velocity. Hence arises a centrifugal force which acts in combination with the force of gravity, and which Newton found to be the 289th part of the force of gravity at the equator, and decreasing as the cosine of the latitude from the equator to the poles. The great predominance of gravity over the centrifugal force prevents the latter from carrying off any bodies from the earth's surface; but the weight of all bodies is diminished by the centrifugal force, so that the weight of any body is greater at the poles than it is at the equator. If we now suppose the waters at the pole to communicate with those at the equator by means of a canal, one branch of which goes from the pole to the centre of the earth, and the other from the centre of the earth to the equator, then the polar branch of the canal will be heavier than the equatorial branch, in consequence of its weight not being diminished by the centrifugal force; and, therefore, in order that the two columns may be in *equilibrio*, the equatorial one must be lengthened. Newton found that the length of the polar must be to that of the equatorial canal as 229 to 230, or that the earth's polar radius must be seventeen miles less than its equatorial radius; that is, that the figure of the earth is an oblate spheroid, formed by the revolution of an ellipse round its lesser axis. Now since the intensity of gravity at any point of the earth's surface increases as the distance of that point from the centre decreases, it follows from this that it increases from the equator to the poles,—a result which he confirmed by the fact, that clocks require to have their pendulums shortened in order to beat true time when carried from Europe towards the equator.

The next subject to which Newton applied the principle of gravity was the tides of the ocean. The philosophers of all ages had recognised the connexion between the phenomena of the tides and the position of the moon. The College of Jesuits at Coimbra, and subsequently Antonio de Dominis and Kepler, distinctly referred the tides to the attraction of the waters of the earth by the moon; but so imperfect was the explanation which was thus given of the phenomena, that Galileo ridiculed the idea of lunar attraction, and substituted for it a fallacious

explanation of his own. That the moon is the principal
cause of the tides is obvious from the well-known fact,
that it is high water at any given place about the time
when she is on the meridian of that place.[*] And that the
sun performs a secondary part in their production may be
proved from the circumstance, that the highest tides take
place when the sun, the moon, and the earth are in the
same straight line,—that is, when the force of the sun
combines with that of the moon; and that the lowest tides
take place when the lines drawn from the sun and moon
to the earth are at right angles to each other,—that is,
when the force of the sun acts in opposition to that of the
moon. The most perplexing phenomenon in the tides of
the ocean, and one which is still a stumbling-block to
persons slightly acquainted with the theory of attraction,
is the existence of high water on the side of the earth
opposite to the moon, as well as on the side next the
moon. To maintain that the attraction of the moon at
the same instant draws the waters of the ocean towards
herself, and also draws them from the earth in an opposite
direction, seems at first sight paradoxical; but the difficulty
vanishes when we consider the earth, or rather the centre
of the earth, and the water on each side of it, as three
distinct bodies placed at different distances from the moon,
and consequently attracted with forces inversely proportional to the squares of their distances. The water nearest
the moon will be much more powerfully attracted than the
centre of the earth, and the centre of the earth more
powerfully than the water farthest from the moon. The

* Perhaps the most striking shape in which to put this connexion is,
that the difference between two consecutive times of high or low water at
any place is always half a lunar day, or the actual difference of time
between the Moon's being on the meridian (above or below the horizon)
at that place. The actual time of high or low water is affected by local
causes, but the difference between consecutive times of those occurrences
shews at once the connexion with the Moon.—EDITOR.

consequence of this must be, that the waters nearest the moon will be drawn away from the centre of the earth, and will consequently rise from their level, while the centre of the earth will be drawn away from the waters opposite the moon, which will, as it were, be left behind, and consequently be in the same situation as if they were raised from the earth in a direction opposite to that in which they are attracted by the moon. Hence the effect of the moon's action upon the earth is to draw its fluid parts into the form of an oblong spheroid, the axis of which passes through the moon. As the action of the sun will produce the very same effect, though in a smaller degree, the tide at any place will depend on the relative position of these two spheroids, and will be always equal either to the sum or to the difference of the effects of the two luminaries. At the time of new and full moon the two spheroids will have their axes coincident, and the height of the tide, which will then be a *spring* one, will be equal to the sum of the elevations produced in each spheroid considered separately, while at the first and third quarters the axes of the spheroids will be at right angles to each other, and the height of the tide, which will then be a *neap* one, will be equal to the difference of the elevations produced in each separate spheroid. By comparing the spring and neap tides, Newton found that the force with which the moon acted upon the waters of the earth was to that with which the sun acted upon them as 4·48 to 1; that the force of the moon produced a tide of 8·63 feet; that of the sun one of 1·93 feet; and both combined, one of 10½ feet,—a result which in the open sea does not deviate much from observation. Having thus ascertained the force of the moon on the waters of our globe, he found that the quantity of matter in the moon was to that in the earth as 1 to 40, and the density of the moon to that of the earth as 11 to 9.

The motions of the moon, so much within the reach of our own observation, presented a fine field for the application of the theory of universal gravitation. The irregularities exhibited in the lunar motions had been known in the time of Hipparchus and Ptolemy. Tycho had discovered the great inequality called the *variation*, amounting to 37', and depending on the alternate acceleration and retardation of the moon in every quarter of a revolution; and he had also ascertained the existence of the annual equation. Of these two inequalities Newton gave a most satisfactory explanation, making the first 36' 10", and the other 11' 51", differing only a few seconds from the numbers adopted by Tobias Mayer in his celebrated Lunar Tables. The force exerted by the sun upon the moon may be always resolved into two forces, one acting in the direction of the line joining the moon and the earth, and consequently tending to increase or diminish the moon's gravity to the earth; and the other in a direction at right angles to this, and consequently tending to accelerate or retard the motion in her orbit. Now, it was found by Newton that this last force was reduced to nothing, or vanished at the syzygies and quadratures, so that at these four points the moon describes areas proportional to the times. The instant, however, that the moon quits these positions, the force under consideration, which we may call the tangential force, begins, and it reaches its maximum in the four octants. The force, therefore, compounded of these two elements of the solar force, or the diagonal of the parallelogram which they form, is no longer directed to the earth's centre, but deviates from it at a maximum by about thirty minutes, and therefore affects the angular motion of the moon, the motion being accelerated in passing from the quadratures to the syzygies, and retarded in passing from the syzygies to the quadratures. Hence the velocity is in its mean state in the

octants, at a maximum in the syzygies, and at a minimum
in the quadratures.

Upon considering the influence of the solar force in
diminishing or increasing the moon's gravity to the earth,
Newton saw that her distance and her periodic time must
from this cause be subject to change, and in this way he
accounted for the annual equation observed by Tycho.
By the application of similar principles, he explained the
cause of the motion of the apsides, or of the greater axis
of the moon's orbit, which has an angular progressive
motion of $3^\circ 4'$ nearly in the course of one lunation; and
he showed that the retrogradation of the nodes, amounting
to $3' 10''$ daily, arose from one of the elements of the
solar force being exerted in the plane of the ecliptic, and
not in the plane of the moon's orbit,—the effect of which
was to draw the moon down towards the plane of the
ecliptic, and thus cause the line of the nodes, or the inter-
section of these two planes, to move in a direction opposite to that of the moon.

The lunar theory thus sketched by Newton, required
for its completion the labours of another century. The
imperfections of the fluxionary calculus prevented him
from explaining the other inequalities of the moon's
motions, and it was reserved for Euler, D'Alembert,
Clairaut, Mayer, and Laplace, to bring the lunar tables to
a high degree of perfection, and to enable the navigator
to determine his longitude at sea with a degree of precision
which the most sanguine astronomer could scarcely have
anticipated.*

* In this place, it is impossible to leave without mention the recent
labours of Hansen and Delaunay, the former of whom presented us with
the lunar tables which are now extensively used, and are far more accurate than those of his predecessors; and the latter was engaged in the
preparation of a set which would have been even superior to Hansen's,
when he unfortunately met with a premature death by drowning on the
5th of August, 1872. The Astronomer Royal, Sir George Airy, is now

By the consideration of the retrograde motion of the moon's nodes, Newton was led to discover the cause of the remarkable phenomenon of the precession of the equinoctial points, which moved 50" annually, and completed the circuit of the heavens in 25,920 years. Kepler had declared himself incapable of assigning any cause for this motion, and we do not believe that any other astronomer had ever made the attempt. From the spheroidal form of the earth, it may be regarded as a sphere with a spheroidal ring surrounding its equator, one-half of the ring being above the plane of the ecliptic, and the other half below it. Considering this excess of matter as a system of satellites adhering to the earth's surface, Newton now saw that the combined actions of the sun and moon upon these satellites tended to produce a retrogradation in the nodes of the circles which they described in their diurnal rotation, and that the sum of all the tendencies being communicated to the whole mass of the planet, ought to produce a slow retrogradation of the equinoctial points. The effect produced by the motion of the sun he found to be 40", and that produced by the action of the moon 10".

Although there could be little doubt that the comets were retained in their orbits by the same laws which regulated the motions of the planets, yet it was not easy to put this opinion to the test of observation. The visibility

working out a scheme by which—availing himself of the labours of his predecessors, particularly of Delaunay,—he hopes to provide the astronomical and nautical world with lunar tables which shall represent the moon's place with the utmost accuracy of modern theory, and arranged in a form better adapted for use than Hansen's. The stupendous work effected by him during the earlier part of his tenure of office at Greenwich, in the complete reduction of the observations of his predecessors, from those of Bradley in 1750, together with their continuation by himself from 1836 with more than previous regularity, have furnished the materials for the late important improvements in lunar tables, and we may well wish for his complete success in making these of the fullest practical use.—EDITOR.

of comets only in a small part of their orbits rendered it difficult to ascertain their distances and periodic times; and as their periods were probably of great length, it was impossible to obtain accurate results by repeated observation. Newton, however, removed this difficulty, by showing how to determine the orbit of a comet, namely, the form and position of the orbit, and the periodic time, by three observations. This method consists of an easy geometrical construction, founded on the supposition that the paths of comets are so nearly parabolic, that the parabola may be used without any sensible error, although he considered it more probable that their orbits are elliptical, and that after a long period they might return. By applying this method to the comet of 1680, he calculated the elements of its orbit, and from the agreement of the computed places with those which were observed, he justly inferred that the motions of comets were regulated by the same laws as those of the planetary bodies. This result was one of great importance; for as the comets enter our system in every possible direction, and at all angles with the ecliptic, and as a great part of their orbits extends far beyond the limits of the solar system, it demonstrated the existence of gravity in spaces beyond the planets, and proved that the law of the inverse ratio of the squares of the distances was true in every possible direction, and at very remote distances from the centre of our system.

Such is a brief view of the leading discoveries which the *Principia* first announced to the world. The grandeur of the subject of which it treats—the beautiful simplicity of the system which it unfolds—the clear and concise reasoning by which that system is explained—and the irresistible evidence by which it is supported, might have insured it the warmest admiration of contemporary mathematicians, and the most welcome reception in all the schools of philosophy throughout Europe. This, however, is not the way in which great truths are generally received. Though the astronomical discoveries of Newton were not assailed by the class of ignorant pretenders who attacked his optical writings, yet they were everywhere resisted by the errors and prejudices which had taken a deep hold even of the strongest minds. The philosophy of Descartes was predominant throughout Europe. Appealing to the imagination more than to the reason, it was quickly received into popular favour, and the same causes which facilitated its introduction, extended its influence, and completed its dominion over the human mind. In explaining all the movements of the heavenly bodies by a system of vortices in a fluid medium diffused through the universe, Descartes had seized upon an analogy of the most alluring and deceitful kind. Those who have seen heavy bodies revolving in the eddies of a whirlpool, or in the gyrations of a vessel of water thrown into a circular motion, had no difficulty in conceiving how the planets might revolve round the sun by analogous movements. The mind instantly grasped at an explanation of so palpable a character, and which required for its development neither the exercise of patient thought, nor the aid of mathematical skill. The talent and perspicuity with which the Cartesian system was expounded, and the show of experiments by which it was illustrated and supported, contributed powerfully to its adoption, while it derived a still higher sanction from the excellent character, and the unaffected piety of its author.

Thus intrenched, as the Cartesian system was, in the strongholds of the human mind, and fortified by its most obstinate prejudices, it was not to be wondered at that the pure and sublime doctrines of the Principia were distrustfully received, and perseveringly resisted. The uninstructed mind could not readily admit the idea, that the great masses of the planets were suspended in empty space,

and retained in their orbits by an invisible influence residing in the sun; and even those philosophers who had been
accustomed to the rigour of true scientific research, and
who possessed sufficient mathematical skill for the examination of the Newtonian doctrines, viewed them at
first as reviving the occult qualities of the ancient physics,
and resisted their introduction with a pertinacity which it
is not easy to explain. Prejudiced, no doubt, in favour
of his own metaphysical views, Leibnitz himself misapprehended the principles of the Newtonian philosophy, and
endeavoured to demonstrate the truths in the Principia
by the application of different principles. Huygens, who
above all other men was qualified to appreciate the new
philosophy, rejected the doctrine of gravitation as existing between the individual particles of matter, and received
it only as an attribute of the planetary masses. John
Bernouilli also, one of the first mathematicians of the
age, opposed the philosophy of Newton. Mairan, in the
early part of his life, was a strenuous defender of the
system of vortices. Cassini and Maraldi were quite
ignorant of the Principia, and occupied themselves with
the most absurd methods of calculating the orbits of
comets long after the Newtonian method had been established on the most impregnable basis; and even Fontenelle,
a man of liberal views and extensive information, continued, throughout the whole of his life, to maintain the
doctrines of Descartes.

The Chevalier Louville of Paris, in his memoir "On
the Construction and Theory of Tables of the Sun," had
applied the doctrine of central force to the motions of
the planets, so early as 1720. S'Gravesande had introduced it into the Dutch universities at a somewhat earlier
period, and Maupertuis, in consequence of a visit which
he paid to England in 1728, became a zealous defender of
it in his Treatise on the Figures of the Celestial Bodies.

But, notwithstanding these and some other examples that might be quoted, we must admit the truth of the remark of Voltaire, that though Newton survived the publication of the Principia more than forty years, yet at the time of his death he had not above twenty followers out of England.

With regard to the progress of the Newtonian philosophy in England, some difference of opinion has been entertained. Professor Playfair gives the following account of it :—"In the universities of England, though the Aristotelian physics had made an obstinate resistance, they had been supplanted by the Cartesian, which became firmly established about the time when their foundation began to be sapped by the general progress of science, and particularly by the discoveries of Newton. For more than thirty years after the publication of these discoveries, the system of vortices kept its ground, and a translation from the French into Latin of the *Physics* of Rohault, a work entirely Cartesian, continued at Cambridge to be the text for philosophical instruction. About the year 1718, a new and more elegant translation of the same book was published by Dr. Samuel Clarke, with the addition of notes, in which that profound and ingenious writer explained the views of Newton on the principal objects of discussion, so that the notes contained virtually a refutation of the text: they did so, however, only virtually, all appearance of argument and controversy being carefully avoided. Whether this escaped the notice of the learned Doctor or not is uncertain, but the new translation, from its better Latinity, and the name of the editor, was readily admitted to all the academical honours which the old one had enjoyed. Thus the stratagem of Dr. Clarke completely succeeded; the tutor might prelect from the text, but the pupil would sometimes look into the notes, and error is never so sure of being exposed, as when the truth

is placed close to it, side by side, without any thing to
alarm prejudice, or awaken from its lethargy the dread of
innovation. Thus, therefore, the Newtonian philosophy
first entered the university of Cambridge under the pro-
tection of the Cartesian." To this passage Professor
Playfair adds the following as a note.

"The Universities of St. Andrews and Edinburgh
were, I believe, the first in Britain where the Newtonian
philosophy was made the subject of the academical pro-
lections. For this distinction they are indebted to James
and David Gregory, the first in some respects the rival,
but both the friends, of Newton. Whiston bewails, in
the anguish of his heart, the difference, in this respect,
between those universities and his own. David Gregory
taught in Edinburgh for several years prior to 1690, when
he removed to Oxford; and Whiston says, 'he had
already caused several of his scholars to keep Acts, as
we call them, upon several branches of the Newtonian
philosophy, while we at Cambridge, poor wretches, were
ignominiously studying the fictitious hypotheses of the
Cartesians.'* I do not, however, mean to say, that from
this date the Cartesian philosophy was expelled from those
universities; the *Physics* of Rohault were still in use as a
text-book, at least occasionally, to a much later period
than this; and a great deal, no doubt, depended on the
character of the individual. Professor Keill introduced
the Newtonian philosophy in his lectures at Oxford in
1697; but the instructions of the tutors, which constitute
the real and efficient system of the university, were not
cast in that mould till long afterwards." Adopting the
same view of the subject, Mr. Dugald Stewart has stated,
"that the philosophy of Newton was publicly taught by
David Gregory at Edinburgh, and by his brother, James

---

* Whiston's *Memoirs of his own Life*, p. 36.

Gregory, at St. Andrews,* before it was able to supplant the vortices of Descartes in that very university of which Newton was a member. It was in the Scottish universities that the philosophy of Locke, as well as that of Newton, was first adopted as a branch of academical education.

Anxious as we should have been to have awarded to Scotland the honour of having first adopted the Newtonian philosophy, yet a regard for historical truth compels us to take a different view of the subject. It is well known that Sir Isaac Newton delivered lectures on his own philosophy from the Lucasian chair before the publication of the *Principia*; and in the very page of Whiston's Life quoted by Professor Playfair, he informs us that he had heard him read such lectures in the public schools, though at that time he did not at all understand them. Newton continued to lecture till 1699, and occasionally, we presume, till 1703, when Whiston became his successor, having been appointed his deputy in 1699. In both these capacities Whiston delivered in the public schools a course of lectures on astronomy, and a course of physico-mathematical lectures, in which the mathematical philosophy of Newton was explained and demonstrated; and both these courses were published—the one in 1707, and the other in 1710—"for the use of the young men in the University." In 1707, the celebrated blind mathematician, Nicholas Saunderson, took up his residence in Christ's College, without being admitted a member of that body. The society not only allotted to him apartments, but gave him the free use of their library. With the concurrence of Whiston he delivered a course of lectures "on the Prin-

---

* Dr. Reid states, that James Gregory, Professor of Philosophy at St. Andrews, printed a Thesis at Edinburgh in 1690, containing twenty-five positions, of which twenty-two were a compend of Newton's *Principia*.

cipia, Optics, and Universal Arithmetic of Newton;" and
the popularity of these lectures was so great, that Sir Isaac
corresponded on the subject of them with their author;
and on the ejection of Whiston from the Lucasian chair
in 1711, Saunderson was appointed his successor. In this
important office he continued to teach the Newtonian
philosophy till the time of his death, which took place in
1739.

But while the Newtonian philosophy was thus regularly
taught in Cambridge, and after the publication of the
Principia, there were not wanting other exertions for
accelerating its progress. About 1694, the celebrated
Dr. Samuel Clarke, while an undergraduate, defended,
in the public schools, a question taken from the Newtonian
philosophy, and his translation of Rohault's Physics, which
contains references in the notes to the Principia, and which
was published in 1697, (and not in 1718, as stated by
Professor Playfair,) shows how early the Cartesian system
was attacked by the disciples of Newton. The author of
the Life of Saunderson informs us, that public exercises
or acts founded on every part of the Newtonian system
were very common about 1707, and so general were such
studies in the university, that the Principia rose to four
times its original price.* One of the most ardent votaries
of the Newtonian philosophy was Dr. Laughton, who had
been tutor in Clare Hall from 1694, and it is probable
that, during the whole, or at least the greater part, of his
tutorship, he had inculcated the same doctrines. In
1709-10, when he was proctor of that college, instead of
appointing a moderator, he discharged the office himself,
and devoted his most active exertions to the promotion of
mathematical knowledge. Previous to this, he had even

* Nichols's Literary Anecdotes, vol. iii., p. 322. Cotes states in his
preface to the second edition of the Principia, that copies of the first
edition could only be obtained at an immense price.

published a paper of questions on the Newtonian philosophy, which appear to have been used as theses for disputations; and such was his ardour and learning that
they powerfully contributed to the popularity of his college.
Between 1706 and 1716, the year of his death, the celebrated Roger Cotes, the friend and disciple of Newton,
filled the Plumian chair of astronomy and experimental
philosophy at Cambridge.   During this period he edited
the second edition of the Principia, which he enriched
with an admirable preface, and thus contributed, by his
writings as well as by his lectures, to advance the philosophy of his master.   About the same time the learned
Dr. Bentley, who first made known the philosophy of his
friend to the readers of general literature, filled the high
office of Master of Trinity College, and could not fail to
have exerted his utmost influence in propagating doctrines
which he so greatly admired.   Had any opposition been
offered to the introduction of the true system of the universe, the talents and influence of these individuals would
have immediately suppressed it, but no such opposition
seems to have been made; and though there may have
been individuals at Cambridge, ignorant of mathematical
science, who adhered to the system of Descartes, and
patronized the study of the Physics of Rohault, yet it is
probable that similar persons existed in the universities of
Edinburgh and St. Andrews; and we cannot regard their
adherence to error as disproving the general fact, that the
philosophy of Newton was quickly introduced into all the
universities of Great Britain.

But while the mathematical principles of the Newtonian
system were ably expounded in our seats of learning, its
physical truths were generally studied, and were explained
and communicated to the public by various lecturers on
experimental philosophy.   The celebrated Locke, who
was incapable of understanding the Principia from his

want of mathematical knowledge, inquired of Huygens if all the mathematical propositions in that work were true. When he was assured that he might depend upon their certainty, he took them for granted, and carefully examined the reasonings and corollaries deduced from them. In this manner he acquired a knowledge of the physical truths in the Principia, and became a firm believer in the discoveries which it contained. In the same manner he studied the treatise on Optics, and made himself master of every part of it which was not mathematical.* From a manuscript of Sir Isaac Newton's, entitled "A Demonstration that the Planets, by their Gravity towards the Sun, may move in Ellipses,"† found among the papers of Mr. Locke, and published by Lord King, it would appear that he himself had been at considerable trouble in explaining to his friend that interesting doctrine. This manuscript is endorsed, "Mr. Newton, March, 1689." It begins with three hypotheses, (the two first being the two laws of motion, and the third the parallelogram of motion,) which introduce the proposition of the proportionality of the areas to the times in motions round an immovable centre of attraction.‡ Three lemmas containing the properties of the ellipse, then prepare the reader for the celebrated proposition, that when a body moves in an ellipse,‖ the attraction is reciprocally as the square of the distance of the body from the focus to which it is attracted. These propositions are demonstrated in a more popular manner than in the Principia; but there can be no doubt that, even in their present modified form, they were beyond the capacity of Mr. Locke.

* Preface to Desaguliers' *Experimental Philosophy*. Dr. Desaguliers states that he was told this anecdote several times by Sir Isaac Newton himself.

† *The Life of John Locke*, Edit. 1830, vol. I., pp. 389-400.

‡ *Principia*, lib. I., prop. I.

‖ *Principia*, lib. I., prop. xi.

Dr. John Keill was the first person who publicly taught natural philosophy "by experiments in a mathematical manner." Desaguliers informs us, that this author "laid down very simple propositions which he proved by experiments, and from these he deduced others more compound, which he still confirmed by experiments, till he had instructed his auditors in the laws of motion, the principles of·hydrostatics and optics, and some of the chief propositions of Sir Isaac Newton concerning light and colours. He began those courses in Oxford, about the year 1704 or 1705, and in that way introduced the love of the Newtonian philosophy." When Dr. Keill left the University, Desaguliers began to teach the Newtonian philosophy by experiments. He commenced his lectures at Harthall, in Oxford, in 1710, and delivered more than a hundred and twenty discourses; and when he went to settle in London, in 1713, he informs us that he found "the Newtonian philosophy generally received among persons of all ranks and professions, and even among the ladies by the help of experiments."

Such were the steps by which the Newtonian philosophy was established in Great Britain. From the time of the publication of the Principia, its mathematical doctrines formed a regular part of academical education, and before twenty years had elapsed, its physical truths were communicated to the public in popular lectures illustrated by experiments, and accommodated to the capacities of those who were not versed in mathematical knowledge. The Cartesian system, though it may have lingered for a while in the recesses of our universities, was soon overturned; and long before his death Newton enjoyed the high satisfaction of seeing his philosophy triumphant in his native land.

In closing our account of the Principia, and in justification of the high eulogium we have pronounced upon it, we may quote the opinions of two of the most distinguished men of the past or the present age.* "It may be justly said," observes Halley, "that so many and so valuable philosophical truths, as are herein discovered, and put past dispute, were never yet owing to the capacity and industry of any one man."—(*Phil. Trans.*, vol. xvi., p. 296.) "The importance and generality of the discoveries," says Laplace, "and the great number of original and profound views, which have been the germ of the most brilliant theories of the philosophers of this century, and all presented with much elegance, will insure to the work on the *Mathematical Principles of Natural Philosophy*, a pre-eminence above all the other productions of human genius."—(*Système du Monde*, Edit. 2de, 1799, p. 336.)

* The temptation to multiply such quotations is great, but we forbear. —EDITOR.

CHAPTER XII.

Previous to the time of Newton, the doctrine of infinite
quantities had been the subject of profound study. The
ancients made the first step in this interesting inquiry by
an ingenious attempt to determine the areas of curves.
Their principles were sound, but their want of an organ-
ized method of operation prevented them from even forming a calculus. The method of exhaustions which they
employed for the purpose consisted in making the curve
a limiting area, to which the inscribed and circumscribed
polygonal figures continually approached by increasing the
number of their sides. The area thus obtained was
obviously the area of the curve. In the case of the parabola, Archimedes showed that its area is two-thirds of its
circumscribing rectangle, or of the product of the ordinate
and the abscissa; and he proved that the superficies of
the sphere was equal to the convex superficies of the circumscribing cylinder, or to four times one of its great

circles, and that the solidity of the sphere was two-thirds
of that of the cylinder. His writings abound in trains of
thought, which are strictly conducted on the principles of
the modern calculus, but in place of this calculus, we have
only an imperfect arithmetic.

The celebrated Pappus of Alexandria followed Archimedes in the same inquiries; and in his demonstration of
the property of the centre of the gravity of a plane figure,
by which may be determined the solid formed by its
revolution, he has shadowed forth the discoveries of
later times.

In his curious tract on Stereometry, published in 1615,
Kepler made some advances in the doctrine of infinitesimals. Prompted to the task by a dispute with the seller
of some casks of wine, he studied the measurement of
solids formed by the revolution of a curve round any line
whatever. In solving some of the simplest of these problems, he conceived a circle to be formed on an infinite
number of triangles having all their vertices in the centre,
and their infinitely small bases in the circumference of the
circle; and by thus rendering familiar the idea of quantities infinitely great and infinitely small, he gave an impulse
to this branch of mathomatics. The failure of Kepler,
too, in solving some of the more difficult of problems
which he himself proposed, roused the attention of geometers, and seems particularly to have attracted the notice of
Cavalieri.

This ingenious mathematician was born at Milan in
1598, and was Professor of Geometry at Bologna. In his
method of Indivisibles, which was published in 1635, he
considered a line as composed of an infinite number of
points, a surface of an infinite number of lines, and a solid
of an infinite number of surfaces; and he lays it down as
an axiom, that the infinite sums of such lines and surfaces
have the same ratio, when compared with the linear or

superficial unit, as the surfaces and solids which are to be
determined.   As it is not true that an infinite number of
infinitely small points can make a line, or an infinite
number of infinitely small lines a surface, Pascal removed
this verbal difficulty by considering a line as composed of
an infinite number of infinitely short lines, a surface as
composed of an infinite number of infinitely narrow paral-
lelograms, and a solid of an infinite number of infinitely
thin solids.   But, independently of this correction, the
conclusions deduced by Cavalieri are rigorously true, and
his method of ascertaining the ratios of areas and solids to
one another, and the theorems which he deduced from it,
may be considered as forming an era in mathematics.

By the application of this method, Roberval and Torri-
celli showed that the area of the cycloid is three times
that of its generating circle, and the former extended the
method of Cavalieri to the case where the powers of the
terms of the arithmetical progression to be summed were
fractional.

In applying the doctrine of infinitely small quantities
to determine the tangents of curves, and the maxima and
minima of their ordinates, both Roberval and Fermat
made a near approach to the invention of Fluxions—so
near indeed that both Lagrange and Laplace* have pronounced the latter to be the true inventor of the differential calculus.   Roberval supposed the point which describes
a curve to be actuated by two motions, by the composition
of which it moves in the direction of a tangent; and had
he possessed the method of fluxions, he could, in every
case, have determined the relative velocities of these motions, which depend on the nature of the curve, and con-

* "On peut regarder Fermat," says Lagrange, "comme le premier inventeur des nouveaux calculs;" and Laplace observes, "Il paraît que
Fermat, le veritable inventeur du calcul differentiel, l'ait envisagé comme
un cas particulier de celui des différences," &c.

sequently the direction of the tangent, which he assumed
to be in a diagonal of a parallelogram whose sides had the
same ratio as the velocities. But as he was able to determine these velocities only in the conic sections, &c., his
ingenious method had but few applications.

The labours of Peter Fermat, a councillor of the parliament of Toulouse, approached still nearer to the fluxionary calculus. In his method of determining the maxima
and minima of the ordinates of curves, he substitutes $x + e$
for the independent variable x in the function which is to
become a maximum, and as these two expressions should
be equal when e becomes infinitely small or 0, he frees
this equation from surds and radicals, and after dividing
the whole by e, e is made $= 0$, and the equation for the
maximum is thus obtained. Upon a similar principle he
founded his method of drawing tangents to curves. But
though these methods thus used by Fermat are in principle
the same with those which connect the theory of tangents
and of maxima and minima with the analytical method of
exhibiting the differential calculus, yet it is a singular
example of national partiality, to consider the inventor of
these methods as the inventor of the method of fluxions.

"One might be led," says Sir J. Herschel, "to suppose
by Laplace's expression that the calculus of finite differences had then already assumed a systematic form, and
that Fermat had actually observed the relation between
the two calculi, and derived the one from the other. The
latter conclusion would scarcely be less correct than the
former. No method can justly be regarded as bearing
any analogy to the differential calculus which does not lay
down a system of rules (no matter on what considerations
founded, by what names called, or by what extraneous
matter enveloped) by means of which the second term of
the development of any function of $x + e$ in powers of e,
can be correctly calculated, 'quæ extendet se,' to use New-

ton's expression, '*citra* ullum molestum calculum in ter-
minis surdis æque ac in integris procedens.' It would be
strange to suppose Fermat or any other in possession of
such a method before any single surd quantity had ever
been developed in a series. But, in point of fact, his
writings present no trace of the kind; and this, though
fatal to his claim, is allowed by both the geometers cited.
Hear Lagrange's candid avowal. ' Il fait disparaitre dans
cette equation,' that of the maximum between x and e,
'les radicaux et les fractions s'il y en a.' Laplace, too,
declares that ' il savoit étendre son calcul aux fonctions
irrationelles en se débarrassant des irrationalités par
l'élévation des radicaux aux puissances.' This is at once
giving up the point in question. It is allowing unequivocally that Fermat in these processes only took a cir-
cuitous route to avoid a difficulty which it is one of the
most express objects of the differential calculus to face
and surmount. The whole claim of the French geometer
arises from a confusion (too often made) of the calculus
and its applications, the means and the end, under the
sweeping head of 'nouveaux calculs' on the one hand,
and an assertion somewhat too unqualified advanced in
the warmth and generality of a preface on the other.*

The discoveries of Fermat were improved and simplified
by Hudde, Huygens, and Barrow; and by the publication
of the *Arithmetic of Infinites* by Dr. Wallis, Savilian
Professor of Geometry at Oxford, mathematicians were
conducted to the very entrance of a new and untrodden
field of discovery. This distinguished author had effected
the quadrature of all curves whose ordinates can be expressed by any direct integral powers; and though he had
extended his conclusions to the cases where the ordinates
are expressed by the inverse or fractional powers, yet he

* Art. MATHEMATICS in the *Edinburgh Encyclopedia*, vol. xiii., p. 865.

failed in its application. Nicolas Mercator (Kaufmann) surmounted the difficulty by which Wallis had been baffled, by the continued division of the numerator by the denominator to infinity, and then applying Wallis's method to the resulting positive powers. In this way he obtained, in 1667, the first general quadrature of the hyperbola, and, at the same time, gave the regular development of a function in series.

In order to obtain the quadrature of the circle, Dr. Wallis considered that if the equation of the curves of which he had given the quadrature were arranged in a series, beginning with the most simple, these areas would form another series. He saw also that the equation of the circle was intermediate between the first and second terms of the first series, or between the equation of a straight line and that of a parabola; and hence he concluded, that by interpolating a term between the first and second terms of the second series, he would obtain the area of the circle. In pursuing this singularly beautiful thought, Dr. Wallis did not succeed in obtaining the indefinite quadrature of the circle, because he did not employ general exponents; but he was led to express the entire area of the circle by a fraction, the numerator and denominator of which are each obtained by the continued multiplication of a certain series of numbers.

Such was the state of this branch of mathematical science, when Newton, at an early age, directed to it the vigour of his mind. At the very beginning of his mathematical studies, when the works of Dr. Wallis fell into his hands, he was led to consider how he could interpolate the general values of the areas in the second series of that mathematician. With this view he investigated the arithmetical law of the co-efficients of the series, and obtained a general method of interpolating not only the series above referred to, but also other series. These were the first steps taken by Newton; and, as he himself informs us, they would have entirely escaped from his memory if he had not, a few weeks before,* found the notes which he made upon the subject. When he had obtained this method, it occurred to him that the very same process was applicable to the ordinates; and, by following out this idea, he discovered the general method of reducing radical quantities composed of several terms into infinite series, and was thus led to the discovery of the celebrated *Binomial Theorem.* He now neglected entirely his methods of interpolation, and employed that theorem alone as the easiest and most direct method for the quadratures of curves, and in the solution of many questions which had not even been attempted by the most skilful mathematicians.

After having applied the Binomial Theorem to the rectification of curves, and to the determination of the surfaces and contents of solids, and the positions of their centres of gravity, he discovered the general principle of deducing the areas of curves from the ordinate, by considering the area as a nascent quantity, increasing by continual fluxion in the proportion of the length of the ordinate, and supposing the abscissa to increase uniformly in proportion to the time. In imitation of Cavalieri, he called the momentary increment of a line a point, though it is not a geometrical point, but an infinitely short line; and the momentary increment of an area or surface he called a line, though it is not a geometrical line, but an infinitely narrow surface. By thus regarding lines as generated by the motion of points, surfaces by the motion of lines, and solids by the motion of surfaces, and by considering that the ordinates, abscissæ, &c., of curves thus formed, vary according to a regular law depending on the

* These facts are mentioned in Newton's letter to Oldenburg, October 24th, 1676.

equation of the curve, he deduces from this equation the velocities with which these quantities are generated; and by the rules of infinite series he obtains the ultimate value of the quantity required. To the velocities with which every line or quantity is generated, Newton gave the name of *Fluxions*, and to the lines or quantities themselves that of *Fluents*. This method constitutes the doctrine of fluxions which Newton had invented previous to 1666, when the breaking out of the plague at Cambridge drove him from that town, and turned his attention to other subjects.

But though Newton had not communicated this great invention to any of his friends, he composed his treatise, entitled, *De Analysi per Equationes Numero Terminorum infinitas*, in which the principle of fluxions and its numerous applications are clearly pointed out. In the month of June, 1669, he communicated this work to Dr. Barrow, who mentioned it in a letter to Mr. Collins, dated the 20th of June, 1669, as " the production of a friend of his residing at Cambridge, who possesses a fine genius for such inquiries." On the 31st July, he transmitted the work to Collins; and having received his approbation of it, he informed him that the name of the author was Newton, a Fellow of his own college, and a young man, who had only two years before taken his degree of M.A. Collins took a copy of this treatise, and returned the original to Dr. Barrow; and this copy having been found among Collins's papers by his friend Mr. William Jones, and compared with the original manuscript borrowed from Newton, it was published with the consent of Newton, in 1711, nearly fifty years after it was written.

Though the discoveries contained in this treatise were not at first given to the world, yet they were made generally known to mathematicians by the correspondence of Collins, who communicated them to James Gregory; to MM. Bertet and Vernon in France; to Slusius in Holland; to Borelli in Italy; and to Strode, Townsend, and Oldenburg, in letters dated between 1669 and 1672.

Hitherto the method of fluxions was known only to the friends of Newton and their correspondents; but, in the first edition of the Principia, which appeared in 1687, he published, for the first time, the fundamental principle of the fluxionary calculus, in the second lemma of the second book. No information, however, is here given respecting the algorithm or notation of the calculus; and it was not until 1693 that it was communicated to the mathematical world in the second volume of Dr. Wallis's works, which were published in that year. This information was extracted from two letters of Newton written in 1692.

About the year 1672, Newton had undertaken to publish an edition of Kinckhuysen's Algebra, with notes and additions. He therefore drew up a treatise, entitled, *A Method of Fluxions*, which he proposed as an introduction to that work; but the fear of being involved in disputes about this new discovery, or perhaps the wish to render it more complete, or to have the sole advantage of employing it in his physical researches, induced him to abandon this design. At a later period of his life he again resolved to give it to the world; but it did not appear till after his death, when it was translated into English, and published in 1736, with a commentary by Mr. John Colson, Professor of Mathematics in Cambridge.*

To the first edition of Newton's Optics, which appeared in 1704, there were added two mathematical treatises, entitled, *Tractatus duo de Speciebus et Magnitudine Figu-*

* Dr. Pemberton informs us that he had prevailed upon Sir Isaac to publish this treatise during his lifetime, and that he had for this purpose examined all the calculations and prepared part of the figures. But as the latter part of the treatise had never been finished, Sir Isaac was about to let him have other papers to supply what was wanting, when his death put a stop to the plan.—Preface to Pemberton's *View of Sir Isaac Newton's Philosophy.*

rarum curvilinearum, the one bearing the title of *Tractatus de Quadraturâ Curvarum*, and the other, *Enumeratio Linearum tertii Ordinis*. The first contains an explanation of the doctrine of fluxions, and of its application to the quadrature of curves; and the second a classification of seventy-two curves of the third order, with an account of their properties. The reason for publishing these two tracts in his Optics, (in the subsequent editions of which they are omitted,) is thus stated in the advertisement:— "In a letter written to M. Leibnitz in the year 1679, and published by Dr. Wallis, I mentioned a method by which I had found some general theorems about squaring curvilinear figures on comparing them with the conic sections, or other the simplest figures with which they might be compared. And some years ago I lent out a manuscript containing such theorems; and having since met with some things copied out of it, I have on this occasion made it public, prefixing to it an introduction, and joining a scholium concerning that method. And I have joined with it another small tract concerning the curvilineal figures of the second kind, which was also written many years ago, and made known to some friends, who have solicited the making it public."

In the year 1707, Mr. Whiston published the algebraical lectures which Newton had, during nine years, delivered at Cambridge, under the title of *Arithmetica Universalis, sive de Compositione et Resolutione Arithmeticâ Liber*. We are not accurately informed how Mr. Whiston obtained possession of this work; but it is stated by one of the editors of the English Edition, that "Mr. Whiston, thinking it a pity that so noble and useful a work should be doomed to a college confinement, obtained leave to make it public." It was soon afterwards translated into English by Mr. Ralphson; and a second edition of it, with improvements by the author, was published at London in 1712, by Dr. Machin, Secretary to the Royal Society. With the view of stimulating mathematicians to write annotations on this admirable work, the celebrated Dutch professor, S'Gravesande,* published a tract, entitled, *Specimen Commentarii in Arithmeticam Universalem;* and Maclaurin's Algebra seems to have been drawn up in consequence of this appeal.

Among the mathematical works of Newton we must not omit to enumerate a small tract entitled, *Methodus Differentialis*, which was published with his consent in 1711. It consists of six propositions, which contain a method of drawing a parabolic curve through any given number of points, and which are useful for constructing tables by the interpolation of series, and for solving problems depending on the quadrature of curves.

Another mathematical treatise of Newton's was published for the first time in 1779, in Dr. Horsley's edition of his works.† It is entitled, *Artis Analyticæ Specimina, vel Geometria Analytica.* In editing this work, which occupies about one hundred and thirty quarto pages, Dr. Horsley used three manuscripts, one of which was in the handwriting of the author; another, written in an unknown hand, was given by Mr. William Jones to the Hon. Charles Cavendish; and a third, copied from the latter by Mr. James Wilson, the editor of Robins's Works, was given to Dr. Horsley by Mr. John Nourse, bookseller to the king. Dr. Horsley has divided it into twelve chapters, which treat of infinite series; of the reduction of affected equations; of the specious resolution of equations; of the doctrine of fluxions; of maxima and minima; of drawing tangents to curves; of the radius of curvature; of some cognate questions; of the quadrature of curves; of the area of curves which are comparable

* Storm van s'Gravesande, Prof. of Mathematics at Leyden.—EDITOR.
† Isaaci Newtoni Opera quæ extant omnia, vol. I, pp. 858-919.

with the conic sections; of the construction of mechanical problems; and of finding the lengths of curves.

In enumerating the mathematical works of our author, we must not overlook his solutions of the celebrated problems proposed by Bernouilli and Leibnitz. In June 1696, John Bernouilli addressed a letter to the most distinguished mathematicians in Europe,[*] challenging them to solve the two following problems :—

1. To determine the curve line connecting two given points which are at different distances from the horizon, and not in the same vertical line, along which a body passing by its own gravity, and beginning to move at the upper point, shall descend to the lower point in the shortest time possible.

2. To find a curve line of this property, that the two segments of a right line drawn from a given point through the curve, being raised to any given power, and then taken together, may make everywhere the same sum.

On the day after he received these problems, Newton addressed to Mr. Charles Montague, the President of the Royal Society, a solution of them both. He announced that the curve required in the first problem must be a cycloid, and he gave a method of determining it. He solved also the second problem, and he showed that by the same method other curves might be found which should cut off three or more segments having the like properties. Leibnitz, who was struck with the beauty of the problem, requested Bernouilli, who had allowed six months for its solution, to extend the period to twelve months. This delay was readily granted; solutions were obtained from Newton, Leibnitz, and the Marquis de l'Hôpital; and although that of Newton was anonymous, yet Bernouilli recognized in it his powerful mind; "*tanquam*," says he, "*ex ungue leonem*," as the lion is known by his claw.

[*] "Acutissimis qui toto orbe florent Mathematicis."

One of the last mathematical efforts of our author was made, with his usual success, in solving a problem which Leibnitz proposed in 1716, in a letter to the Abbé Conti, " for the purpose," as he expressed it, " of feeling the pulse of the English analysis." The object of this problem was to determine the curve which should cut at right angles an infinity of curves of a given nature, but expressible by the same equation. Newton received this problem about five o'clock in the afternoon, as he was returning from the Mint; and though the problem was extremely difficult, and he himself much fatigued with business, yet he finished the solution of it before he went to bed.

Such is a brief account of the mathematical writings of Sir Isaac Newton, not one of which were voluntarily communicated to the world by himself. The publication of his Universal Arithmetic is said to have been a breach of confidence on the part of Whiston; and, however this may be, it was an unfinished work, never designed for the public. The publication of his *Quadrature of Curves*, and of his *Enumeration of Curve Lines*, was rendered necessary, in consequence of plagiarisms from the manuscripts of them which he had lent to his friends; and the rest of his analytical writings did not appear till after his death. It is not easy to penetrate into the motives by which this great man was on these occasions actuated. If his object was to keep possession of his discoveries till he had brought them to a higher degree of perfection, we may approve of the propriety, though we cannot admire the prudence of such a step. If he wished to retain to himself his own methods, in order that he alone might have the advantage of them in prosecuting his physical inquiries, we cannot reconcile so selfish a measure with that openness and generosity of character which marked the whole of his life. If he withheld his labours from the world in order to avoid the disputes and contentions to which they might give rise, he adopted the very worst method of securing his tranquillity.   That this was the leading motive under which he acted, there is little reason to doubt.   The early delay in the publication of his method of fluxions, after the breaking out of the plague at Cambridge, was probably owing to his not having completed the whole of his design ; but no apology can be made for the imprudence of withholding it any longer from the public—an imprudence which is the more inexplicable, as he was repeatedly urged by Wallis, Halley, and his other friends, to present it to the world.   Had he published this noble discovery previous to 1673, when his great rival had made but little progress in those studies which led him to the same method, he would have secured to himself the undivided honour of the invention, and Leibnitz could have aspired to no other fame but that of an improver of the doctrine of fluxions. But he unfortunately acted otherwise.   He announced to his friends that he possessed a method of great generality and power; he communicated to them a general account of its principles and applications; and the information which was thus conveyed, might have directed the attention of mathematicians to subjects to which they would not have otherwise applied their powers.   The discoveries which he had previously made, were made subsequently by others; and Leibnitz, instead of appearing on the theatre of science as the disciple and the follower of Newton, stood forth with all the dignity of a second inventor; and, by the early publication of his discoveries, had nearly placed himself on the throne which Newton was destined to ascend.

It would be inconsistent with the nature of this work to enter into a detailed history of the dispute between Newton and Leibnitz respecting the invention of fluxions. A brief and general account of it, however, is indispensable.

In the beginning of 1673, when Leibnitz came to London in the suite of the Duke of Hanover, he became acquainted with the great men who then adorned the capital of England. Among these was Oldenburg, a countryman of his own, who was at that time secretary to the Royal Society. About the beginning of March in the same year, Leibnitz went to Paris, where, with the assistance of Huygens, he devoted himself to the study of the higher geometry. In the month of July he renewed his correspondence with Oldenburg, and he communicated to him some of the discoveries which he had made relative to series, particularly the series for a circular arc in terms of the tangent. Oldenburg informed him in return of the discoveries on series which had been made by Newton and Gregory; and in 1676 Newton communicated to him, through Oldenburg, a letter of fifteen closely-printed quarto pages, containing many of his analytical discoveries, and stating, that he possessed a general method of drawing tangents which he thought it necessary to conceal in two sentences of transposed characters. In this letter neither the method of fluxions nor any of its principles are communicated; but the superiority of the method over all others is so fully described, that Leibnitz could scarcely fail to discover that Newton possessed that secret of which geometers had so long been in quest.

Had Leibnitz at the time of receiving this letter been entirely ignorant of his own differential method, the information thus conveyed to him by Newton could not fail to stimulate his curiosity, and excite his mightiest efforts to obtain possession of so great a secret. That this new method was intimately connected with the subject of series was clearly indicated by Newton; and as Leibnitz was deeply versed in this branch of analysis, it is far from improbable that a mind of such strength and acuteness might attain his object by direct investigation. That this was the case

N

may be inferred from his letter to Oldenburg (to be communicated to Newton) of the 21st of June, 1677, where he
mentions that he had for some time been in possession of
a method of drawing tangents more general than that of
Slusius, namely, by the differences of ordinates. He then
proceeds with the utmost frankness to explain this method,
which was no other than the differential calculus. He
describes the algorithm which he had adopted, the formation of differential equations, and the application of the
calculus to various geometrical and analytical questions. No
answer seems to have been returned to this letter either by
Newton or Oldenburg; and with the exception of a short
letter from Leibnitz to Oldenburg, dated 12th of July,
1677, no further correspondence seems to have taken place.
This no doubt, arose from the death of Oldenburg in the
month of August, 1678;* and the two rival geometers
having, through him, become acquainted with each other's
labours, were left to pursue them with all the ardour
which the importance of the subject could not fail to
inspire.

In the hands of Leibnitz the differential calculus made
rapid progress. In the *Acta Eruditorum*, which appeared
at Leipsic in October, 1684, he described its algorithm in
the same manner as he had done in his letter to Oldenburg,

* Henry Oldenburg, whose name is so intimately associated with the
history of Newton's discoveries, was born at Bremen, and was consul
from that town to London during the usurpation of Cromwell. Having
lost his office, and being compelled to seek the means of subsistence, he
became tutor to an English nobleman, whom he accompanied to Oxford
in 1650. During his residence in that city he became acquainted with
the philosophers who established the Royal Society; and upon the death
of William Crown, the first secretary, he was appointed, in 1663, joint
secretary along with Mr. Wilkins. He kept up an extensive correspondence with the philosophers of all nations, and he was the author of
several papers in the Philosophical Transactions, and of some works which
have not acquired much celebrity. He died at Charlton, near Greenwich,
in August, 1678.

and pointed out its application to the drawing of tangents,
and the determination of maxima and minima; remarking
that these were but the beginnings of a much more sublime
geometry, applicable to the most difficult and beautiful
problems even of mixed mathematics which, without his
differential calculus, or one *similar* to it, could not be
treated with equal facility. The suppression of Newton's
name in this reference to a *similar* calculus, which was
obviously that of Newton, referred to by him in the letters
of 1676, was the first false step in the fluxionary controversy, and may be regarded as its commencement. In
June, 1686, Leibnitz resumed the subject; and when
Newton had not published a single word upon fluxions,
and had not even made known his notation, the differential
calculus was making rapid advances on the Continent,
and in the hands of James and John Bernouilli had proved
the means of solving some of the most important and
difficult problems.

The silence of Newton was at last broken, and in the
second lemma of the second book of the Principia, he
explained the fundamental principle of the fluxionary calculus. His explanation, which occupied only three pages,
was terminated with the following scholium: "In a
correspondence which took place about ten years ago
between that very skilful geometer, G. G. Leibnitz,* and
myself, I announced to him that I possessed a method of
determining maxima and minima, of drawing tangents, and
of performing similar operations, which was equally applicable to rational and irrational quantities, and concealed
the same in transposed letters involving this sentence,
*(datâ equatione quotcunque fluentes quantitates involvente,
fluxiones invenire et vice versâ.)* This illustrious man
replied that he also had fallen on a method of the same

---

* Gottfried Wilhelm Leibnitz. The initials above are in Latin.—EDITOR.

kind, and he communicated his method, which scarcely
differed from my own except in the notation [*and in the
idea of the generation of quantities.] The fundamental
principle of both is contained in this lemma."

This celebrated scholium, which has been the subject of
such angry discussion, has, in our opinion, been much
misapprehended. While M. Biot considers it as "eter-
nalizing the rights of Leibnitz by recognizing them in the
Principia," Professor Playfair regards it as containing
"a highly favourable opinion on the subject of the discoveries of Leibnitz." To us it appears to be nothing
more than the simple statement of the fact, that the method
communicated by Leibnitz was nearly the same as his
own; and this much he might have said, whether he
believed that Leibnitz had seen the fluxionary calculus
among the papers of Collins, or was the independent
inventor of his own. It is more than probable, indeed,
that when Newton wrote this scholium he regarded Leibnitz as a second inventor; but when he found that Leibnitz
and his friends had shown a willingness to believe, and
had even ventured to throw out the suspicion, that he
himself had borrowed the doctrine of fluxions from the
differential calculus, he seems to have altered the opinion
which he had formed of his rival, and to have been
willing in his turn to retort the charge.

This change of opinion was brought about by a series
of circumstances over which he had no control. M.
Nicolas Facio de Duillier, a Swiss mathematician, resident
in London, communicated to the Royal Society, in 1699,
a paper on the line of quickest descent, which contains
the following observations: "Compelled by the evidence
of facts, I hold Newton to have been the first inventor of
this calculus, and the earliest by several years; and

* These words in brackets are in the second edition, but not in
the first.

whether Leibnitz, the *second inventor*, has borrowed any thing from the other, I would prefer to my own judgment that of those who have seen the letters and other copies of the same manuscripts of Newton." This imprudent remark, which by no means amounts to a charge of plagiarism, for Leibnitz is actually designated the *second inventor*, may be considered as showing that the English mathematicians had been cherishing suspicions unfavourable to Leibnitz, and there can be no doubt that a feeling had long prevailed that this mathematician either had or might have seen, among the papers of Collins, the "*Analysis per Equationes*," &c., which contained the principles of the fluxionary method. Leibnitz replied to the remark of Duillier with much good feeling. He appealed to the facts as exhibited in his correspondence with Oldenburg; he referred to Newton's scholium as a testimony in his favour; and without disputing or acknowledging the priority of Newton's claim, he asserted his own right to the invention of the differential calculus. Facio transmitted a reply to the Leipsic Acts; but the editor refused to insert it. The dispute, therefore, terminated, and the feelings of the contending parties continued for some time in a state of repose, though ready to break out on the slightest provocation.

When Newton's Optics appeared in 1704, accompanied by his Treatise on the Quadrature of Curves, and his Enumeration of Lines of the third Order, the editor of the Leipsic Acts (whom Newton supposed to be Leibnitz himself) took occasion to review the first of these tracts. After giving an imperfect analysis of its contents, he compared the method of fluxions with the differential calculus; and, in a sentence of some ambiguity, he states that Newton employed fluxions in place of the differences of Leibnitz, and made use of them in his Principia in the same manner as Honoratus Fabri, in his Synopsis of

Geometry, had substituted progressive motion in place of the indivisibles of Cavalieri.* As Fabri, therefore, was not the inventor of the method which is here referred to, but borrowed it from Cavalieri, and only changed the mode of its expression, there can be no doubt that the artful insinuation contained in the above passage was intended to convey the impression that Newton had *stolen* his method of fluxions from Leibnitz. The indirect character of this attack, in place of mitigating its severity, renders it doubly odious; and we are persuaded that no candid reader can peruse the passage without a strong conviction that it justifies, in the fullest manner, the indignant feelings which it excited among the English mathematicians. If Leibnitz was the author of the review, or if he was in any way a party to it, he merited the full measure of rebuke which was dealt out to him by the friends of Newton, and deserved those severe reprisals which doubtless embittered the rest of his days. He who dared to accuse a man like Newton, or indeed any man holding a fair character in society, with the odious crime of plagiarism, placed himself without the pale of the ordinary courtesies of life, and deserved to have the same charge thrown back upon himself. The man who conceives his fellow to be capable of such intellectual felony, avows the possibility of himself committing it, and almost substantiates the weakest evidence of the worst accusers.

Dr. Keill, as the representative of Newton's friends, could not brook this concealed attack upon his countryman.

* As this passage is of essential importance in this controversy, we shall give it in the original:—*" Pro differentiis igitur Leibnitianis D. Newtonus adhibet, sempergue adhibuit, fluxiones, quæ sunt quàm proximè ut fluentium augmenta, æqualibus temporis particulis quàm minimis genita; iisque tum in suis Principiis Naturalis Philosophiæ Mathematicis, tum in aliis postea edltis, eleganter est usus; quemadmodum et Honoratus Fabrius in sua Synopsi Geometricâ, motuum progressus Cavallerianæ methodo substituit."*

In a letter on the Laws of Centripetal Forces, addressed
to Halley, and printed in the Philosophical Transactions
for 1708, he maintained that Newton was "beyond all
doubt" the first inventor of fluxions. He referred, for a
direct proof of this, to his letters published by Wallis;
and he asserted "that the same calculus was afterwards
published by Leibnitz, the name and the mode of notation
being changed." If the reader is disposed to consider this
passage as retorting the charge of plagiarism upon Leibnitz,
he will readily admit that the mode of its expression is
neither so coarse nor so insidious as that which is used by
the writer in the Leipsic Acta. In a letter to Hans Sloane,
dated March, 1711, Leibnitz complained to the Royal
Society of the treatment he had received. He expressed his
conviction that Keill had erred more from rashness of judgment than from any improper motive, and said that he did
not regard the accusation as a calumny; and he requested
that the Society would oblige Keill to disown publicly the
injurious sense which his words might bear. When this
letter was read to the Society, Keill justified himself to
Sir Isaac Newton and the other members by showing them
the obnoxious review of the Quadrature of Curves in the
Leipsic Acta. They all agreed in attaching the same
injurious meaning to the passage which we formerly
quoted, and authorized Keill to explain and defend his
statement. He accordingly addressed a letter to Sir Hans
Sloane, which was read at the Society on the 24th of May,
1711, and a copy of which was ordered to be sent to
Leibnitz. In this letter, which is one of considerable
length, he declares that he never meant to state that
Leibnitz knew either the name of Newton's method or the
form of notation, and that the real meaning of the passage
was, "that Newton was the first inventor of fluxions or
of the differential calculus, and that he had given, in two
letters to Oldenburg, and which he had transmitted to

Leibnitz, indications of it sufficiently intelligible to an acute
mind, from which Leibnitz derived, or at least might
derive, the principles of his calculus."

The charge of plagiarism which Leibnitz thought was
implied in the former letter of his antagonist is here greatly
modified, if not altogether denied. Keill expresses only
an *opinion* that the letter *seen* by Leibnitz contained
intelligible indications of the fluxionary calculus. Even
if this opinion were correct, it is no proof that Leibnitz
either saw these indications or availed himself of them;
or if he did perceive them, it might have been in consequence of his having previously been in possession of the
differential calculus, or having enjoyed some distant view
of it. Leibnitz should, therefore, have allowed the dispute
to terminate here; for no ingenuity on his part and no
additional facts could affect an opinion which any other
person as well as Keill was entitled to maintain.

Leibnitz, however, took a different view of the subject,
and wrote a letter to Sir Hans Sloane, dated December
19th, 1711, which excited new feelings, and involved him
in new embarrassments. Insensible to the mitigation
which had been kindly impressed upon the supposed charge
against his honour, he alleges that Keill had attacked his
candour and sincerity more openly than before;—that he
acted without any authority from Sir Isaac Newton, who
was the party interested;—and that it was in vain to
justify his proceedings by referring to the provocation in
the Leipsic Acts, because, in that Journal, *no injustice had
been done to any party, but every one had received what
was his due.* He branded Keill with the odious appellation
of an upstart, and one little acquainted with the circumstances of the case; * he called upon the Society to silence

* "Homine docto, sed novo, et parum perito rerum antecedentium
cognitione."

his vain and unjust clamours,* which, he believed, were disapproved by Newton himself, who was well acquainted with the facts, and who, he was persuaded, would willingly give his opinion on the matter.

This unfortunate letter was doubtless the cause of all the rancour and controversy which so speedily followed, and it placed his antagonist in a new and a more favourable position. It may be correct, though few will admit it, that Keill's second letter was more injurious than the first; but it was not true that Keill acted without the authority of Newton, because Keill's letter was approved of and transmitted by the Royal Society, of which Newton was the president, and therefore became the act of that body. The obnoxious part, however, of Leibnitz's letter consisted in his appropriating to himself the opinions of the reviewer in the Leipsic Acts, by declaring, that, in a review which charged Newton with plagiarism, every person had received what was his due. The whole character of the controversy was now changed: Leibnitz places himself in the position of the party who had first disturbed the tranquillity of science by maligning its most distinguished ornament; and the Royal Society was imperiously called upon to throw all the light they could upon a transaction which had exposed their venerable president to so false a charge. The Society, too, had become a party to the question, by their approbation and transmission of Keill's second letter, and were on that account alone bound to vindicate the step which they had taken.

When the letter of Leibnitz, therefore, was read, Keill appealed to the registers of the Society for the proofs of what he had advanced; Sir Isaac also expressed his displeasure at the obnoxious passage in the *Acta Eruditorum*, and at the defence of it by Leibnitz, and he left it

* " Vanæ et injustæ vociferationes."

to the Society to act as they thought proper. A committee was therefore appointed on the 11th of March, consisting of Dr. Arbuthnot, Mr. Hill, Dr. Halley, Mr. Jones, Mr. Machin, and Mr. Burnet, who were instructed to examine the ancient registers of the Society, to inquire into the dispute, and to produce such documents as they should find, together with their own opinions on the subject. On the 24th of April the committee produced the following Report:—

" We have consulted the letters and letter-books in the custody of the Royal Society, and those found among the papers of Mr. John Collins, dated between the years 1669 and 1677, inclusive; and showed them to such as knew and avouched the hands of Mr. Barrow, Mr. Collins, Mr. Oldenburg, and Mr. Leibnitz; and compared those of Mr. Gregory with one another, and with copies of some of them taken in the hand of Mr. Collins; and have extracted from them what relates to the matter referred to us; all which extracts herewith delivered to you, we believe to be genuine and authentic. And by these letters and papers we find,

"I. Mr. Leibnitz was in London in the beginning of the year 1673; and went thence, in or about March, to Paris; where he kept a correspondence with Mr. Collins, by means of Mr. Oldenburg, till about September, 1676, and then returned by London and Amsterdam to Hanover: and that Mr. Collins was very free in communicating to able mathematicians what he had received from Mr. Newton and Mr. Gregory.

"II. That when Mr. Leibnitz was the first time in London, he contended for the invention of another differential method, properly so called; and, notwithstanding that he was shown by Dr. Pell that it was Newton's method, persisted in maintaining it to be his own invention, by reason that he had found it by himself, without knowing what Newton had done before, and had much improved it. And we find no mention of his having any other differential method than Newton's, before his letter of the 21st of June, 1677, which was a year after a copy of Mr. Newton's letter, of the 10th of December, 1672, had been sent to Paris to be communicated to him; and above four years after Mr. Collins began to communicate that letter to his correspondent; in which letter the method of fluxions was sufficiently described to any intelligent person.

"III. That by Mr. Newton's letter of the 13th of June, 1676, it appears that he had the method of fluxions above five years before the writing of that letter. And by his 'Analysis per Æquationes Numero Terminorum infinitas,' communicated by Dr. Barrow to Mr. Collins in July, 1669, we find that he had invented the method before that time.

"IV. That the differential method is one and the same with the method of fluxions, excepting the name and mode of notation; Mr. Leibnitz calling these quantities differences, which Mr. Newton calls moments or fluxions; and marking them with the letter d,—a mark not used by Mr. Newton. And therefore we take the proper question to be, not who invented this or that method, but who was the first inventor of the method. And we believe, that those who have reputed Mr. Leibnitz the first inventor knew little or nothing of his correspondence with Mr. Collins and Mr. Oldenburg long before; nor of Mr. Newton's having that method above fifteen years before Mr. Leibnitz began to publish it in the *Acta Eruditorum* of Leipsic.

"For which reason we reckon Mr. Newton the first inventor; and are of opinion, that Mr. Keill, in asserting the same, has been no ways injurious to Mr. Leibnitz. And we submit to the judgment of the Society, whether the extract and papers now presented to you, together with what is extant to the same purpose in Dr. Wallis's third volume, may not deserve to be made public."

This report being read, the Society unanimously ordered the collection of letters and manuscripts to be printed, and appointed Dr. Halley, Mr. Jones, and Mr. Machin, to superintend the press. Complete copies of it under the title of *Commercium Epistolicum D. Johannis Collins et aliorum de Analysi promota*, were laid before the Society on the 8th of January, 1713; and Sir Isaac Newton, as president, ordered a copy to be delivered to each person of the Committee appointed for that purpose, to examine it before its publication.

Leibnitz received information of the appearance of the Commercium Epistolicum when he was at Vienna; and "being satisfied," as he expresses it, "that it must contain *malicious falsehoods*, I did not think proper to send for it by post, but wrote to M. Bernouilli to give me his sentiments. Mr. Bernouilli wrote me a letter, dated at Basle, June 7th, 1713, in which he said, *that it appeared probable that Sir Isaac Newton had formed his calculus after having seen mine.*"[*] This letter was published by a friend of Leibnitz, with reflections, in a loose sheet entitled *Charta Volans*, and dated July 29th, 1713. It was widely circulated without either the name of the author, printer, or place of publication, and was communicated to the *Journal Littéraire* by another friend of Leibnitz, who added remarks of his own, and stated, that when Newton published the Principia in 1687, *he did not understand the true differential method; and that he took his fluxions from Leibnitz.*

In this state of the controversy, Mr. Chamberlayne conceived the design of reconciling the two distinguished

* Letter to Count Bothman in Des Malzeaux's *Recueil de diverses Pièces*, tom. li., pp. 44, 45.

philosophers; and in a letter, dated April 28th, 1714, [*]
he addressed himself to Leibnitz, who was still at Vienna.
In replying to this letter, Leibnitz declared that he had
given no occasion for the dispute; "that Newton procured
a book to be published which was written purposely to
discredit him, and sent it to Germany, &c., as in the name
of the Society;" and he stated, "*that there was room to
doubt whether Newton knew his invention before he had it
of him.*" Mr. Chamberlayne communicated this letter to
Sir Isaac Newton, who replied, that Leibnitz had attacked
his reputation in 1705, by intimating that he had borrowed
from him the method of fluxions; that if Mr. C. could
point out to him any thing in which he had injured Mr.
Leibnitz, he would endeavour to give him satisfaction;
that he would not retract things which he knew to be true;
and that he believed that the Royal Society had done no
injustice by the publication of the Commercium Epistolicum.

The Royal Society, having learned that Leibnitz complained of their having condemned him unheard, inserted
a declaration in their journals on the 20th of May, 1714,
that they did not pretend that the report of their committee
should pass for a decision of the Society. Mr. Chamberlayne sent a copy of this to Leibnitz, along with Sir Isaac's
letter, and Dr. Keill's answer to the papers inserted in the
Journal Littéraire. After perusing these documents, M.
Leibnitz replied, "that Sir Isaac's letter was written with
very little civility; that he was not in an humour to put
himself in a passion against such people; that there were
other letters among those of Oldenburg and Collins which
should have been published; and that on his return to
Hanover, he would be able to publish a Commercium
Epistolicum which would be of service to the history of

* See Des Maizeaux, tom. II, p. 116.

learning." When this letter was read to the Royal
Society, Sir Isaac remarked, that the last part of it
injuriously accused the Society of having made a partial
selection of papers for the Commercium Epistolicum;
that he did not interfere in any way in the publication of
that work; and had even withheld from the committee two
letters, one from Leibnitz in 1693, and another from Wallis
in 1695, which were highly favourable to his cause. He
stated that he did not think it right for M. Leibnitz himself to publish such a paper as he proposed; but that, if
he had letters to produce in his favour, they might be published in the Philosophical Transactions, or in Germany.

About this time the Abbé Conti, a noble Venetian,
came to England. He was a correspondent of Leibnitz;
who, in a letter received by Conti soon after his arrival,[*]
entered upon his dispute with Newton. He charged the
English " with wishing to pass for almost the only in-
ventors." He declared " that Bernouilli has judged rightly
in 'saying that Newton did not possess before him the
infinitesimal characteristic and algorithm." He remarked
that Newton preceded him only in series; and confessed
that during his second visit to England, " Collins showed
him part of his correspondence," or, as he afterwards
expressed it, he saw " some of the letters of Newton at
Mr. Collins's." He then attacked Sir Isaac's philosophy,
particularly his opinions about gravity, vacuum, and the
intervention of God for the preservation of His creatures;
and he accused him of reviving the occult qualities of the
schools. But the most remarkable passage in this letter is
the following: " I am a great friend of experimental
philosophy, but Newton deviates much from it *when he
pretends that all matter is heavy*, or that each particle of
matter attracts every other particle."

---

* Written in November or December, 1715.

The above letter to the Abbé Conti was generally shown in London, and came to be much talked of at court, in consequence of Leibnitz having been privy councillor to the Elector of Hanover when that prince ascended the throne of England. Many persons of distinction, and particularly the Abbé Conti, urged Newton to reply to Leibnitz's letter, but he resisted all their solicitations. One day, however, King George I. inquired when Sir Isaac Newton's answer to Leibnitz would appear; and when Sir Isaac heard this, he addressed a long reply to the Abbé Conti, dated February 26th (O. S.,) 1715–16. This letter, written with dignified severity, is a triumphant refutation of the allegations of his adversary; and the following passage deserves to be quoted, as connected with that branch of the dispute which relates to Leibnitz's having seen part of Newton's letters to Mr. Collins. "He complains of the committee of the Royal Society, as if they had acted partially in omitting what made against me; but he fails in proving the accusation. For he instances in a paragraph concerning my ignorance, pretending that they omitted it; and yet you will find it in the Commercium Epistolicum, p. 547, lines 2, 3, and I am not ashamed of it. He saith that he saw this paragraph in the hands of Mr. Collins when he was in London the second time, that is, in October, 1676. It is in my letter of the 24th of October, 1676, and therefore he then saw that letter. And in that and some other letters writ before that time, I described my method of fluxions; and in the same letter I described also two general methods of series, one of which is now claimed from me by Mr. Leibnitz." The letter concludes with the following paragraph: "But as he has lately attacked me with an accusation which amounts to plagiary; if he goes on to accuse me, it lies upon him by the laws of all nations to prove his accusations, on pain of being accounted guilty of calumny.

He hath hitherto written letters to his correspondents full of affirmations, complaints, and reflections, without proving any thing.　But he is the aggressor, and it lies upon him to prove the charge."

In transmitting this letter to Leibnitz, the Abbé Conti informed him that he himself had read with great attention, and without the least prejudice, the Commercium Epistolicum, and the little piece* that contains the extract; that he had also seen at the Royal Society the original papers of the Commercium Epistolicum, and some other original pieces relating to it.　"From all this," says he, "I infer, that, if all the digressions are cut off, the only point is, whether Sir Isaac Newton had the method of fluxions or infinitesimals before you, or whether you had it before him. You published it first, it is true, but you have owned also that Sir Isaac Newton had given many hints of it in his letters to Mr. Oldenburg and others.　This is proved very largely in the Commercium, and in the extract of it. What answer do you give?　This is still wanting to the public, in order to form an exact judgment of the affair." The Abbé adds, that Mr. Leibnitz's own friends waited for his answer with great impatience, and that they thought he could not dispense with answering, if not Dr. Keill, at least Sir Isaac Newton himself, who had given him a defiance in express terms.

Leibnitz was not long in complying with this request. He addressed a letter to the Abbé Conti on the 9th of April, 1716, but he sent it through M. Rémond de Montmort at Paris, to communicate it to others.　When it was received by the Abbé Conti, Newton wrote observations upon it, which were communicated only to some of his friends, and which, while they placed his defence on the

---

* This is the *Recensio Commercii Epistolici*, or review of it, which was first published in the *Phil. Trans.*, 1715.

most impregnable basis, at the same time threw much light on the early history of his mathematical discoveries.

The death of Leibnitz on the 14th of November, 1716, put an end to this controversy; and Newton some time afterwards published the correspondence with the Abbé Conti, which had hitherto been only privately circulated among the friends of the disputants.*

In the year 1725, a new edition of the Commercium Epistolicum was published, with notes, a general review of it, and a preface of some length. A question has arisen respecting the authorship of the review and the preface, some ascribing it to Keill, and others to Newton. From similarity of style, but chiefly on the authority of Dr. James Wilson, the friend of Pemberton, Professor De Morgan had made it highly probable that both the review and the preface were written by Newton. Of the correctness of this opinion, I have found ample evidence in the manuscripts at Hurstbourne Park; and it is due to historical truth to state that Newton supplied all the materials for the Commercium Epistolicum, and that, though Keill was its editor, and the committee of the Royal Society the authors of the Report, Newton was virtually responsible for its contents.

The share which Newton took in the fluxionary controversy, either directly or through Dr. Keill, who did nothing without his approbation, and the mass of papers

* M. Biot remarks that the animosity of Newton was not calmed by the death of Leibnitz; *for he had no sooner heard of it than he caused to be printed two manuscript letters of Leibnitz, written in the preceding year, accompanying them with a very bitter refutation,* (en les accompagnant d'une réfutation très-amère.) Who on reading this sentence would not suppose that the " bitter " refutation was written after Leibnitz's death? The animosity could not be shown by the simple *publication* of the letters. It could reside only in the *bitterness* of the *refutation.* The implied charge is untrue; the refutation was written before Leibnitz's death, and consequently Newton showed no animosity over the grave of his rival; nor in our opinion can he justly be charged with any even before his death.

which he has left behind him on the subject, show the
great anxiety which he felt not only to be considered the
first inventor of the calculus, but the only inventor who
had a right to the reputation which it gave. He firmly
believed not only that Leibnitz might have derived the
differential calculus from the papers actually communicated
to him, but that he did derive it from that source, or from
his own ideas either oral or written, which were in circula-
tion at the time of his visit to London. That these were
the views of Newton, and, we may add, of all his friends in
England, is evident from the new form given to the celebrated scholium in the third edition of the *Principia*,
which appeared in 1725, under the editorship of Pemberton.
The reference to Leibnitz and his method was wholly
omitted, and replaced by a quotation from a letter to
Collins, in December 1672, containing, or supposed to
contain, the germ of Fluxions. This step was perhaps
unwise. The statement in the first two editions granted
nothing to Leibnitz, and, even if it had, the truth which
it embodied was not cancelled by its omission from the
third; but viewing the matter as Newton did, we think
he was justified in omitting the scholium. He had stated
it, as he himself has said, as a mere historical fact that
Leibnitz had sent him a method which was similar to his
own; and when he found that the German mathematician
had regarded this simple statement as a recognition of his
independent discovery of the calculus, he was not only
entitled, but constrained, to cancel a passage which had
been so erroneously interpreted, and so improperly used.

In viewing this controversy, at this distance of time,
when the passions of the individual combatants have been
allayed, and national jealousies extinguished, it is not
difficult to form a correct estimate of the conduct and
claims of the two rival analysts. By the unanimous
verdict of all nations, it has been decided that Newton

invented fluxions at least ten years before Leibnitz. Some
of the letters of Newton which bore reference to this
great discovery were perused by the German mathematician; but there is no evidence whatever that he borrowed
his differential calculus from these letters. Newton was
therefore the *first* inventor, and Leibnitz the *second*. It
was impossible that the former could have been a plagiarist;
but it was possible for the latter. Had the letters of
Newton contained even stronger indications than they do
of the new calculus, no evidence short of proof could have
justified any allegation against Leibnitz's honour. The
talents which he displayed in the improvement of the
calculus showed that he was capable of inventing it; and
his character stood sufficiently high to repel every suspicion
of his integrity. But if it would have been criminal to
charge Leibnitz with plagiarism, what must we think of
those who dared to accuse Newton of borrowing his fluxions
from Leibnitz? This odious accusation was made by
Leibnitz himself, and by Bernouilli; and we have seen
that the former repeated it again and again, as if his own
good name rested on the destruction of that of his rival.
It was this charge against Newton that gave rise to the
attack of Keill, and the publication of the Commercium
Epistolicum; and, notwithstanding this high provocation,
the Committee of the Royal Society contented themselves
with asserting Newton's priority without retorting the
charge of plagiarism upon his rival.

Although an attempt has been recently made to place
the conduct of Leibnitz on the same level with that of
Newton, yet the circumstances of the case will by no
means justify such a comparison. The conduct of Newton
was at all times dignified and just. He knew his rights,
and he boldly claimed them. Conscious of his integrity,
he spurned with indignation the charge of plagiarism with
which an ungenerous rival had so insidiously loaded him;

and if there was one step in his frank and unhesitating
procedure which posterity can blame, it is his omission, in
the third edition of the Principia, of the reference to the
differential calculus of Leibnitz.   This omission, however,
was perfectly just.   The scholium which he had left out
was a mere historical statement of the fact, that the German
mathematician had sent him a method which was the same
as his own; and when he found that this simple assertion
had been held by Leibnitz and others as a recognition of
his independent claim to the invention, he was bound either
to omit it altogether, or to enter into explanations which
might have involved him in a new controversy.

The conduct of Leibnitz was not marked with the same
noble lineaments.   That he was the aggressor is universally
allowed.   That he first dared to breathe the charge of
plagiarism against Newton, and that he often referred to
it, has been sufficiently apparent; and when arguments
failed him, he had recourse to threats,—declaring that he
would publish another Commercium Epistolicum, though
he had no appropriate letters to produce.   All this is now
matter of history; and we may find some apology for it
in his excited feelings, and in the insinuations which were
occasionally thrown out against the originality of his discovery; but for other parts of his conduct we seek in
vain for an excuse.   When he assailed the philosophy of
Newton in his letters to the Abbé Conti, he exhibited
perhaps only the petty feelings of a rival; but when he
dared to calumniate that great man in his correspondence
with the Princess of Wales, by whom he was respected
and beloved; when he ventured to denounce his philosophy
as physically false, and as dangerous to religion; and
when he founded these accusations on passages in the
Principia and the *Optics*, glowing with all the fervour of
genuine piety, he cast a blot upon his name, which all his
talents as a philosopher will never be able to efface.

CHAPTER XIII.

James II. attacks the Privileges of the University of Cambridge—Newton chosen one of the Delegates to resist this Encroachment—He is elected a Member of the Convention Parliament—Burning of his Manuscripts—His supposed Derangement of Mind—View taken of this by Foreign Philosophers—His correspondence with Mr. Pepys and Mr. Locke at the Time of his Illness—Mr. Millington's Letter to Mr. Pepys on the Subject of Newton's Illness—Refutation of the Statement that he laboured under mental Derangement.

In the early chapters of this work, we have brought down the personal history of Newton to the year 1675, when he was permitted by the Crown to retain the Lucasian chair without going into orders. During the first twenty years of his residence at Cambridge, from 1667 to 1687, when the *Principia* was published, he was wholly occupied with those profound researches of which we have given a full account, and tradition has preserved but a few anecdotes of a life so quiet and unvaried.

But, towards the end of that time, an event occurred which drew Newton from the seclusion of his studies, and placed him in a noble position on the theatre of public life. Desirous of re-establishing the Catholic religion in its former supremacy, King James II. had begun to assail the rights and privileges of his Protestant subjects. Among other illegal acts, he, in February, 1687, sent his letter of mandamus to the University of Cambridge, to

order Father Francis, an ignorant monk of the Benedictine
order, to be received as Master of Arts, and to enjoy all
the privileges of this degree, without taking the oaths of
allegiance and supremacy. The University speedily perceived the consequences which might arise from such a
measure. Independently of the infringement of their rights
which such an order involved, it was obvious that the
highest interests of the University were endangered, and
that Roman Catholics might soon become a majority in
the Convocation. They therefore unanimously refused to
listen to the royal order, and they did this with a firmness
of purpose which irritated the despotic court. The King
reiterated his commands, and accompanied them with the
severest threatenings in case of disobedience. The
Catholics were not idle in supporting the views of the
sovereign. The honorary degree of M. A., which conveys
no civil rights to its possessor, having been formerly given
to the Secretary of the Ambassador from Morocco, it was
triumphantly urged that the University of Cambridge had
a greater regard for a Mahometan than for a Roman
Catholic, and was more obsequious to the Ambassador
from Morocco than to their own lawful sovereign.
Though this reasoning might impose upon the ignorant, it
produced little effect upon the members of the University.
A few weak-minded individuals, however, were disposed
to yield a reluctant consent to the royal wishes. They
proposed to confer the degree, and at the same time to
resolve that it should not in future be regarded as a precedent. To this it was replied, that the very act of submission in one case would be a stronger argument for
continuing the practice than any such resolution would be
against its repetition. The University accordingly remained firm in their original decision. The vice-chancellor
was summoned before the ecclesiastical commission to
answer for this act of contempt. Newton was among the

number of those who resisted the wishes of the court, and
he was consequently chosen one of the nine delegates who
were appointed to defend the independence of the University These delegates appeared before the High Court.
They maintained that not a single precedent could be
found to justify so extraordinary a measure; and they
showed that Charles II. had, under similar circumstances,
been pleased to withdraw his mandamus.   This representation had its full weight, and the King was induced to
abandon his design.*

The part which Newton had taken in this affair, and
the high character which he now held in the scientific
world, induced his friends to propose him as a Member of
Parliament for the University. He was accordingly elected
in 1688, though by a very narrow majority,† and he sat
in the Convention Parliament till its dissolution. In the
year 1688 and 1689 Newton was absent from Cambridge
during the greater part of the year, owing, we presume, to
his attendance in Parliament; but it appears from the
books of the University, that, from 1690 to 1695, he was
seldom absent, and must therefore have renounced his
Parliamentary duties.

During his stay in London he had no doubt experienced
the unsuitableness of his income to the new circumstances
in which he was placed, and it is probable that this was
the cause of the limitation of his residence to Cambridge.
His income was certainly very confined, and but little
suited to the generosity of his disposition. Demands were
doubtless made upon it by some of his less wealthy

* See Burnet's *History of his own Times*, vol. l., p. 607. Lond., 1724.

† The other candidates were Sir Robert Sawyer and Mr. Finch, and
the votes stood thus:

Sir Robert Sawyer,	125	} elected.
Mr. Newton,	122	
Mr. Finch,	117	

relatives; and there is reason to think that he himself, as well as his influential friends, had been looking forward to some act of liberality on the part of the Government.

An event, however, occurred, which will ever form an epoch in his history; and it is a singular circumstance that this incident has been for more than a century unknown to his own countrymen, and has been accidentally brought to light by the examination of the manuscripts of Huygens. This event has been magnified into a temporary aberration of mind, which is said to have arisen from a cause scarcely adequate to its production.

While he was attending divine service in a winter morning, he had left in his study a favourite little dog called Diamond. Upon returning from chapel he found that it had overturned a lighted taper on his desk, which set fire to several papers on which he had recorded the results of some optical experiments. These papers are said to have contained the labours of many years, and it has been stated that, when Newton perceived the magnitude of his loss, he exclaimed, "O Diamond, Diamond, little do you know the mischief you have done me." It is a curious circumstance that Newton never refers to the experiments which he is said to have lost on this occasion; and his nephew, Mr. Conduit, makes no allusion to the event itself. The distress, however, which it occasioned is said to have been so deep as to affect even the powers of his understanding.

This extraordinary effect was first communicated to the world in the Life of Newton by M. Biot, who received the following account of it from the celebrated M. Van Swinden.

"There is among the manuscripts of the celebrated Huygens, a small journal in folio, in which he used to note down different occurrences. It is side ζ, No. 8, p. 112, in the catalogue of the library of Leyden. The following extract is written by Huygens himself, with whose hand-writing I am well acquainted, having had occasion to peruse several of his manuscripts and autograph letters. '*On the 29th of May*, 1694, *M. Colin,* * *a Scotsman, informed me that, eighteen months ago, the illustrious geometer, Isaac Newton, had become insane, either in consequence of his too intense application to his studies, or from excessive grief at having lost, by fire, his chemical laboratory and several manuscripts. When he came to the Archbishop of Cambridge,† he made some observations which indicated an alienation of mind. He was immediately taken care of by his friends, who confined him to his house, and applied remedies, by means of which he had now so far recovered his health that he began to understand the Principia.'* Huygens mentioned this circumstance to Leibnitz in a letter, dated 8th of June, 1694, to which Leibnitz replies in a letter dated tho 23rd, 'I am very glad that I received information of the cure of Mr. Newton at the same time that I first heard of his illness, which doubtless must have been very alarming. It is to men like you and him, Sir, that I wish a long life, and much health, more than others, whose loss, comparatively speaking, would not be so great.'*

The first publication of the preceding statement produced a strong sensation among the friends and admirers of Newton. They could not easily believe in the prostration of that intellectual strength which had unbarred the strong-holds of the universe. The unbroken equanimity of

---

* This M. Colin was probably a young Bachelor of Arts whom Newton seems afterwards to have employed in some of his calculations. These bachelors were distinguished by the title of Dominus, and it was usual to translate this word, and to call them, *Sir.* In a letter from Newton to Flamsteed, dated Cambridge, June 29th, 1695, is the following passage: " I want not your calculations, but your observations only; for besides myself, and my servant, Sir Collins (whom I can employ for a little money, which I value not) tells me that he can calculate an eclipse, and work truly."   † Conjectured to mean *Canterbury.*

Newton's mind, the purity of his moral character, his temperate and abstemious life, his ardent and unaffected piety, and the weakness of his imaginative powers, all indicated a mind which was not likely to be overset by any affliction to which it could be exposed. The loss of a few experimental records could never have disturbed the equilibrium of a mind like his. If they were the records of discoveries, the discoveries, themselves indestructible, would have been afterwards given to the world. If they were merely the details of experimental results, a little time could have easily reproduced them. Had these records contained the first fruits of early genius, of obscure talent, on which fame had not yet shed its rays, we might have supposed that the first blight of such early ambition would have unsettled the stability of an untried mind. But Newton was satiated with fame. His mightiest discoveries were completed, and diffused over all Europe, and he must have felt himself placed on the loftiest pinnacle of earthly ambition. The incredulity which such views could not fail to encourage, was increased by the novelty of the information. No English biographer had ever alluded to such an event. History and tradition were equally silent; and it was not easy to believe that the Lucasian Professor of Mathematics at Cambridge, a member of the English Parliament, and the first philosopher in Europe, could have lost his reason without the dreadful fact being known to his own countrymen.

But if the friends of Newton were surprised by the nature of the intelligence, they were distressed at the view which was taken of it by foreign philosophers. While one maintained that the intellectual exertions of Newton had terminated with the publication of the Principia, and that the derangement of his mind was the cause of his abandoning the sciences, others indirectly questioned the sincerity of his religious views, and ascribed to the aberration

of his mind those theological pursuits which gilded his
declining age. "But the fact," says M. Biot, "of the
derangement of his intellect, whatever may have been the
cause of it, will explain why, after the publication of the
Principia in 1687, Newton, though only forty-five years
old, never more published a new work on any branch of
science; but contented himself with giving to the world
those which he had composed long before that epoch,
confining himself to the completion of those parts which
might require development. We may also remark, that
even these developments appear always to be derived from
experiments and observations formerly made, such as the
additions to the second edition of the Principia, published
in 1713, the experiments on thick plates, those on defrac-
tion, and the chemical queries placed at the end of the Optics
in 1704; for in giving an account of these experiments
Newton distinctly says, that they were taken from ancient
manuscripts which he had formerly composed; and he
adds, that though he felt the necessity of extending them,
or rendering them more perfect, he was not able to resolve
to do this, these matters being no longer in his way. Thus
it appears that though he had recovered his health sufficiently to understand all his researches, and even in some
cases to make additions to them, and useful alterations, as
appears from the second edition of the Principia, for
which he kept up a very active mathematical correspondence with Mr. Cotes, yet he did not wish to undertake
new labours in those departments of science where he had
done so much, and where he so distinctly saw what remained to be done." Under the influence of the same
opinion, M. Biot finds "it extremely probable that his
dissertation on the scale of heat was written before the
fire in his laboratory;" he describes Newton's conduct
about the Longitude Bill as exhibiting an inexplicable
timidity of mind, and as "so puerile for so solemn an

occasion, that it might lead to the strangest conclusions, particularly if we refer it to the fatal accident which befell him in 1695."

The celebrated Marquis de la Place viewed the illness of Newton in a light still more painful to his friends. He maintained that he never recovered the vigour of his intellect, and he was persuaded that Newton's theological inquiries did not commence till after that afflicting epoch of his life. He even commissioned Professor Gautier of Geneva to make inquiries on this subject during his visit to England, as if it concerned the interests of truth and justice to show that Newton became a Christian and a theological writer, only after the decay of his strength and the eclipse of his reason.

Such having been the consequence of the disclosure of Newton's illness by the manuscript of Huygens, I felt it to be a sacred duty to the memory of that great man, to the feelings of his countrymen, and to the interests of Christianity itself, to inquire into the nature and history of that indisposition which seems to have been so much misrepresented and misapplied. From the ignorance of so extraordinary an event which has prevailed for such a long period in England, it might have been urged with some plausibility, that Huygens had mistaken the real import of the information that was conveyed to him; or that the Scotchman from whom he received it had propagated an idle and a groundless rumour. But we are fortunately not confined to this very reasonable mode of defence. There exists at Cambridge a manuscript journal written by Mr. Abraham de la Pryme, who was a student in the university while Newton was a Fellow of Trinity. This manuscript is entitled "*Ephemeris Vitæ*, or Diary of my own Life, containing an account likewise of the most observable and remarkable things that I have taken notice of from my youth up hitherto." Mr. de la Pryme was born in 1671, and begins the diary in 1685. This manuscript is in the possession of his collateral descendant, George Pryme, Esq., Professor of Political Economy at Cambridge, to whom I have been indebted for the following extract.

"1692, *February 3rd.*—What I heard to-day I must relate. There is one Mr. Newton, (whom I have very oft seen,) Fellow of Trinity College, that is mighty famous for his learning, being a most excellent mathematician, philosopher, divine, &c. He has been Fellow of the Royal Society these many years; and amongst other very learned books and tracts he's written one upon the mathematical principles of philosophy, which has got him a mighty name, he having received, especially from Scotland, abundance of congratulatory letters for the same; but of all the books that he ever wrote, there was one of colours and light, established upon thousands of experiments which he had been twenty years in making, and which had cost him many hundreds of pounds. This book, which he valued so much, and which was so much talked of, had the ill luck to perish and be utterly lost just when the learned author was almost at putting a conclusion at the same, after this manner: In a winter's morning, leaving it amongst his other papers on his study table whilst he went to chapel, the candle, which he had unfortunately left burning there too, catched hold by some means of other papers, and they fired the aforesaid book, and utterly consumed it and several other valuable writings; and, which is most wonderful, did no further mischief. But when Mr. Newton came from chapel, and had seen what was done, every one thought he would have run mad; he was so troubled thereat, that he was not himself for a month after. A long account of this his system of light and colours you may find in the Transactions of the Royal

Society, which he had sent up to them long before this sad mischance happened unto him."

From this extract we are enabled to fix the approximate date of the accident by which Newton lost his papers. It must have been previous to the 3rd of January, 1692, a month before the date of the extract; but if we fix it by the dates in Huygens's manuscript, we should place it about the 29th of November, 1692, eighteen months previous to the conversation between Collins and Huygens. The manner in which Mr. Pryme refers to Newton's state of mind is that which is used every day when we speak of the loss of tranquillity which arises from the ordinary afflictions of life; and the meaning of the passage amounts to nothing more than that Newton was very much troubled by the destruction of his papers, and did not recover his serenity, and return to his usual occupations, for a month. The very phrase, that every person thought he would have run mad, is in itself a proof that no such effect was produced; and, whatever degree of indisposition may be implied in the phrase, " he was not himself for a month after," we are entitled to infer that one month was the period of its duration, and that previous to the 3rd of February, 1692, the date of Mr. Pryme's memorandum, " Newton was himself again."

These facts and dates cannot be reconciled with those in Huygens's manuscript. It appears from that document, that, so late as May, 1694, Newton had only *so far* recovered his health as *to begin to again understand the Principia*. His supposed malady, therefore, was in force from the 3rd of January, 1692, till the month of May, 1694,—a period of more than two years. Now, it is a most important circumstance, which M. Biot ought to have known, that in *the very middle of this period*, Newton wrote his four celebrated letters to Dr. Bentley on the Existence of a Deity,—letters which evince a power of thought and a serenity of mind absolutely incompatible even with the slightest obscuration of his faculties. No man can peruse these letters without the conviction that their author then possessed the full vigour of his reason, and was capable of understanding the most profound parts of his writings. The first of these letters was written on the 10th of December, 1692, the second on the 17th of January, 1693, the third on the 25th of February, and the fourth on the 11th* of February, 1693. His mind was, therefore, strong and vigorous on these four occasions; and as the letters were written at the express request of Dr. Bentley, who had been appointed to deliver the lecture founded by Mr. Boyle for vindicating the fundamental principles of natural and revealed religion, we must consider such a request as showing his opinion of the strength and freshness of his friend's mental powers.

In 1692, Newton, at the request of Dr. Wallis, transmitted to him the first proposition of his book on quadratures, with examples of it in first, second, and third fluxions.† These examples were written in consequence of an application from his friend; and the author of the review of the Commercium Epistolicum, in which this fact is quoted, draws the conclusion, that he had not at that time forgotten his mode of second fluxions. It appears, also, from the second book of the Optics, ‡ that in the month of June, 1692, he had been occupied with the subject of halos, and had made accurate observations both on the colours and the diameters of the rings in a halo which he had then seen around the sun.

But though these facts stand in direct contradiction to

* They are thus dated in Horsley's edition of Newton's Works, the *fourth* letter having an earlier date than the *third.*

† See *Newtoni Opera,* tom. iv., p. 480, and *Wallisii Opera,* 1693, tom. II., pp. 891-896.

‡ Optics, part iv., obs. 18.

the statement recorded by Huygens, the reader will be naturally anxious to know the real nature and extent of the indisposition to which it refers. The following letters,[*] written by Newton himself, Mr. Pepys, Secretary to the Admiralty, and Mr. Millington of Magdalene College, Cambridge, will throw much light upon the subject.

Newton, as will be presently seen, had fallen into a bad state of health some time in 1692, in consequence of which both his sleep and his appetite were greatly affected. About the middle of September, 1693, he had been kept awake for five nights by this nervous disorder, and in this condition he wrote the following letter to Mr. Pepys :—

"Sir,   *Sept.* 13*th*, 1693.

"Some time after Mr. Millington had delivered your message, he pressed me to see you the next time I went to London. I was averse; but upon his pressing consented, before I considered what I did; for I am extremely troubled at the embroilment I am in, and have neither ate nor slept well this twelvemonth, nor have my former consistency of mind. I never designed to get any thing by your interest, nor by King James's favour, but am now sensible that I must withdraw from your acquaintance, and see neither you nor the rest of my friends any more, if I may but leave them quietly. I beg your pardon for saying I would see you again, and rest your most humble and most obedient servant,

"Is. Newton."

From this letter we learn, on his own authority, that his complaint had lasted for a twelvemonth, and that during that twelvemonth he neither ate nor slept well, nor enjoyed his former *consistency of mind.* It is not easy to

* For these letters I have been indebted to the kindness of Lord Braybrooke.

understand exactly what is meant by not enjoying his former consistency of mind; but whatever be its import, it is obvious that he must have been in a state of mind so sound as to enable him to compose the four letters to Bentley, all of which were written during the twelvemonth here referred to.

On the receipt of this letter, his friend Mr. Pepys seems to have written to Mr. Millington of Magdalene College, to inquire after Mr. Newton's health; but the inquiry having been made in a vague manner, an answer equally vague was returned. Mr. Pepys, however, who seems to have been deeply anxious about Newton's health, addressed the following more explicit letter to his friend Mr. Millington :—

"Sir, *September 26th*, 1693.

"After acknowledging your many old favours, give me leave to do it a little more particularly upon occasion of the new one conveyed to me by my nephew, Jackson. Though at the same time, I must acknowledge myself not at the ease I would be glad to be at in reference to the excellent Mr. Newton; concerning whom (methinks) your answer labours under the same kind of restraint which (to tell you the truth) my asking did. For I was loth at first dash to tell you, that I had lately received a letter from him so surprising to me for the inconsistency of every part of it, as to be put into great disorder by it, from the concernment I have for him, lest it should arise from that, which of all mankind I should least dread from him and most lament for,—I mean a discomposure in head, or mind, or both. Let me therefore beg you, Sir, having now told you the true ground of the trouble I lately gave you, to let me know the very truth of the matter, as far at least as comes within your knowledge. For I own too great an esteem for Mr. Newton, as for a public good, to

P

be able to let any doubt in me of this kind concerning him lie a moment uncleared, where I can have any hopes of helping it. I am, with great truth and respect, dear Sir, your most humble and most affectionate servant,

"S. Pepys."

To this letter Mr. Millington made the following reply:—

Coll. Magd., Camb.,

"Honor'd Sir,   Sept. the 30th, 1693.

"Coming home from a journey on the 28th instant at night, I met with your letter which you were pleased to honour me with of the 26th. I am much troubled I was not at home in time for the post, that I might as soon as possible put you out of your generous payne that you are in for the worthy Mr. Newton. I was, I must confess, very much surprised at the inquiry you were pleased to make by your nephew about the message that Mr. Newton made the ground of his letter to you, for I was very sure I never either received from you or delivered to him any such; and therefore I went immediately to wayt upon him, with a design to discourse him about the matter, but he was out of town, and since I have not seen him, till upon the 28th I met him at Huntingdon, where, upon his own accord, and before I had time to ask him any question, he told me that he had writt to you a very odd letter, at which he was much concerned; added, that it was in a distemper that much seized his head, and that kept him awake for above five nights together, which upon occasion he desired I would represent to you, and beg your pardon, he being very much ashamed he should be so rude to a person for whom he hath so great an honour. He is now very well, and, though I fear he is under some small degree of melancholy, yet I think there is no reason to suspect it hath at all touched his understanding, and I hope never will; and so I am sure all ought to wish that love learning or the honour of our nation, *which it is a sign how much it is looked after, when such a person as Mr. Newton lyes so neglected by those in power.* And thus, honoured Sir, I have made you acquainted with all I know of the cause of such inconsistencys in the letter of so excellent a person; and I hope it will remove the doubts and fears you are, with so much compassion and publickness of spirit, pleased to entertain about Mr. Newton; but if I should have been wanting in any thing tending to the more full satisfaction, I shall, upon the least notice, endeavour to amend it with all gratitude and truth. Honored Sir, your most faithfull and most obedient servant,

"JOH. MILLINGTON."

Mr. Pepys was perfectly satisfied with this answer, as appears from the following letter :—

" SIR,   *October 3rd,* 1693.

"You have delivered me from a fear that indeed gave me much trouble, and from my very heart I thank you for it, an evil to Mr. Newton being what every good man must feel for his own sake as well as his. God grant it may stopp here. And for the kind reflection hee has since made upon his letter to mee, I dare not take upon mee to judge what answer I should make him to it, or whether any or no; and therefore pray that you will bee pleased either to bestow on mee what directions you see fitt for my own guidance towards him in it, or to say to him in my name, but your own pleasure, whatever you think may be most welcome to him upon it, and most expressive of my regard and affectionate esteem of him, and concernment for him.—
•   •   •   •   •   •   Dear Sir, your most humble and most faithful servant,

"S. PEPYS."

It does not appear from the memoirs of Mr. Pepys, whether he ever returned any answer to the letter of Mr. Newton, which occasioned this correspondence; but we find that in less than two months after the date of the preceding letter an opportunity occurred of introducing to him a Mr. Smith, who wished to have his opinion on some problem in the doctrine of chances. This letter from Pepys is dated November 22nd, 1693. Newton replied to it on the 26th of November, and wrote to Pepys again on the 16th of the following month; and in both these letters he enters fully into the discussion of the mathematical question which had been submitted to his judgment.*

It is obvious from Newton's letter to Mr. Pepys, that the subject of his receiving some favour from the government had been a matter of anxiety with himself, and of discussion among his friends.† Mr. Millington was no doubt referring to this anxiety, when he represents him as an honour to the nation, and expresses his surprise "that such a person should *lye so neglected by those in pow-r.*" And we find the same subject distinctly referred to in two letters written by Newton to Mr. Locke during the preceding year. In one of these, dated January 26th, 1691-2, he says, "Being fully convinced that Mr. Montague, upon an old grudge which I thought had been worn out, is false to me, I have done with him, and intend to sit still, unless my Lord Monmouth be still my friend." Mr. Locke seems to have assured him of the continued friendship of this nobleman, and Newton, still referring to the same

---

* These three letters have been published by Lord Braybrooke in the Life and Correspondence of Mr. Pepys.

† This anxiety will be understood from the fact, that by an order of council, dated January 28th, 1674-5, Mr. Newton was excused from making the usual payments of one shilling per week "on account of his low circumstances, as he represented."

topic, in a letter dated February 16th, 1691-2, remarks, "I am very glad Lord Monmouth is still my friend, but intend not to give his Lordship and you any farther trouble. My inclinations are to sit still." In a later letter to Mr. Locke, dated September, 1693, and given in the following page, he asks his pardon for saying or thinking that there was a design to sell him an office. In these letters Newton no doubt referred to some appointment in London which he was solicitous to obtain, and which Mr. Montague and his other friends may have failed in procuring. This opinion is confirmed by the letter of Mr. Montague, announcing to him his appointment to the wardenship of the Mint, in which he says that he is very glad he can *at last* give him good proof of his friendship.

In the same month in which Newton wrote to Mr. Pepys, we find him in correspondence with Mr. Locke. Displeased with his opinions respecting innate ideas, he had rashly stated that they struck at the root of all morality; and that he regarded the author of such doctrines as a Hobbist. Upon re-considering these opinions, he addressed the following remarkable letter to Locke, written three days after his letter to Mr. Pepys, and consequently during the illness under which he then laboured;—

"Sir,

"Being of opinion that you endeavoured to embroil me with women, and by other means, I was so much affected with it, as that when one told me you were sickly and would not live, I answered, 'twere better if you were dead. I desire you to forgive me this uncharitableness; for I am now satisfied that what you have done is just, and I beg your pardon for my having hard thoughts of you for it, and for representing that you struck at the root of morality, in a principle you laid in your book of ideas, and designed to pursue in another book, and that I took you for a Hobbist.* I beg your pardon also for saying or thinking that there was a design to sell me an office, or to embroil me. I am your most humble and unfortunate servant, "IS. NEWTON.

"*At the Bull, in Shoreditch, London,*
　　September 16th, 1693."

To this letter Locke returned the following answer, so nobly distinguished by philosophical magnanimity and Christian charity:—

"SIR, *Oates, October 5th, 1693.*

"I HAVE been, ever since I first knew you, so entirely and sincerely your friend, and thought you so much mine, that I could not have believed what you tell me of yourself, had I had it from any body else. And though I cannot but be mightily troubled that you should have had so many wrong and unjust thoughts of me, yet next to the return of good offices, such as from a sincere good will I have ever done you, I receive your acknowledgment of the contrary as the kindest thing you have done me, since it gives me hopes I have not lost a friend I so much valued. After what your letter expresses, I shall not need to say any thing to justify myself to you. I shall always think your own reflection on my carriage, both to you and all mankind, will sufficiently do that. Instead of that, give me leave to assure you that I am more ready to forgive you than you can be to desire it; and I do it so freely and fully, that I wish for nothing more than the opportunity to convince you that I truly love and esteem you, and that

---

* The system of Hobbes was at this time very prevalent. According to Dr. Bentley, "the taverns and coffee-houses, nay, Westminster Hall, and the very churches, were full of it," and he was convinced from personal observation, that "not one English infidel in a hundred was other than a Hobbist."—Monk's *Life of Bentley*, p. 81.

I have the same good will for you as if nothing of this had happened.   To confirm this to you more fully, I should be glad to meet you any where, and the rather, because the conclusion of your letter makes me apprehend it would not be wholly useless to you.   But whether you think it fit or not, I leave wholly to you.   I shall always be ready to serve you to my utmost, in any way you shall like, and shall only need your commands or permission to do it.

"My book is going to press for a second edition; and, though I can answer for the design with which I write it, yet, since you have so opportunely given me notice of what you have said of it, I should take it as a favour if you would point out to me the places that gave occasion to that censure, that, by explaining myself better, I may avoid being mistaken by others, or unawares doing the least prejudice to truth or virtue.   I am sure you are so much a friend to them both, that were you none to me, I could expect this from you.   But I cannot doubt but you would do a great deal more than this for my sake, who, after all, have all the concern of a friend for you, wish you extremely well, and am, without compliment," &c.*

To this letter Newton made the following reply:—

"Sir,
"The last winter, by sleeping too often by my fire, I got an ill habit of sleeping; and a distemper, which this summer has been epidemical, put me farther out of order, so that when I wrote to you, I had not slept an hour a night for a fortnight together, and for five days together not a wink.   I remember I wrote to you, but what I said of your book I remember not.   If you please to send me

---

* The draught of this letter is indorsed "J. L. to I. Newton."

a transcript of that passage, I will give you an account of
it if I can.   I am your most humble servant,

" Is. NEWTON.

"Cambridge, October 5th, 1693."

Although the first of these letters evinces the existence
of a nervous irritability which could not fail to arise from
want of appetite and of rest, yet it is obvious that its
author was in the full possession of his mental powers.
The answer of Mr. Locke, indeed, is written upon that
supposition ; and it deserves to be remarked, that Mr.
Dugald Stewart, who first published a portion of these
letters, never imagines for a moment that Newton was
labouring under any mental alienation.

The erroneous opinion that Newton devoted his attention to theology only in the latter part of his life, may be
considered as deriving some countenance from the fact,
that the celebrated general Scholium, at the end of the
second edition of the Principia, published in 1713, did not
appear in the first edition of that work.   This argument
has been ably controverted by the late Dr. J. C. Gregory
of Edinburgh, on the authority of a manuscript of Newton, which seems to have been transmitted to his ancestor,
Dr. David Gregory, between the years 1687 and 1698.
This manuscript, which consists of twelve folio pages in
Newton's handwriting, contains, in the form of additions,
and scholia to some propositions in the third book of the
Principia, an account of the opinions of the ancient philosophers on gravitation and motion, and on natural theology, with various quotations from their works.   Attached
to this manuscript are three very curious paragraphs.
The two first appear to have formed the original draught of
the general Scholium already referred to ; and the third
relates to the subject of an ethereal medium, respecting
which he maintains an opinion diametrically opposite to

that which he afterwards published at the end of his
Optics.* The first paragraph expresses nearly the same
ideas as some sentences in the scholium beginning " Deus
summus est ens æternum, infinitum, absolutè perfectum;" †
and it is remarkable that the second paragraph is found
only in the third edition of the Principia, which appeared
in 1726, the year before Newton's death.

In the middle of the year 1694, about the time when
our author is said to have been only beginning again to
understand his own *Principia*, we find him occupied with
the difficult and profound subject of the lunar theory. In
order to procure observations for verifying the equations
which he had deduced from the theory of gravity, he paid
a visit to Flamsteed, at the Royal Observatory of Greenwich, on the 1st of September, 1694, when he received
from him a series of lunar observations. On the 7th of
October he wrote to Flamsteed that, after comparing the
observations with his "conception," he was satisfied that by
both together the moon's theory might be reduced to the
exactness of two or three minutes. He wrote him again
on the 1st of November, and the correspondence was con-

* Dr. Gregory concludes his account of this manuscript, which he has
kindly permitted me to read, in the following words:—" I do not know
whether it is true, as stated by Huygens, ' Newtonum incidisse in phren-
itim;' but I think every gentleman who examines this manuscript will
be of opinion that he must have thoroughly recovered from his phrenitis
before he wrote either the Commentary on the Opinions of the Ancients,
or the Sketch of his own Theological and Philosophical Opinions which
it contains."

† This paragraph is as follows:—" Deum esse ens summè perfectum
concedunt omnes.   Entis autem summè perfecti Idea est ut sit substan-
tia, una, simplex, individibilia, viva et vivifica, ubique semper necessariò
existens, summè intelligens omnia, liberè volens bona, voluntate efficiens
possibilia, effectibus nobilioribus similitudinem propriam quantùm fieri
potest communicans, omnia in se continens tanquam eorum principium
et locus, omnia per præsentiam substantialem cernens et regens, et cum
rebus omnibus, secundùm leges accuratas, ut naturæ totius fundamentum
et causa constanter coöperans, nisi ubi aliter agere bonum est."

tinued till 1698, Newton making constant application for observations to compare with his theory of the planetary motions; while Flamsteed, not sufficiently aware of the importance of the inquiry, received his requests as if they were idle intrusions in which the interests of science were but slightly concerned.*

In reviewing the details which we have now given respecting the health and occupations of Newton from the beginning of 1692 till 1695, it is impossible to draw any other conclusion than that he possessed a sound mind, and was perfectly capable of carrying on his mathematical, his metaphysical, and his astronomical inquiries. His friend and admirer, Mr. Pepys, residing within fifty miles of Cambridge, had never heard of his being attacked with any illness till he inferred it from the letter to himself written in September, 1693. Mr. Millington, who lived in the same University, had been equally unacquainted with any such attack; and, after a personal interview with Newton, for the express purpose of ascertaining the state

* It is not desirable here to enter in detail into the unfortunate dispute between these two great men. The late Mr. Baily, as is well known, in his *Account of Flamsteed*, published in 1835, gives a view of it which reflects very strongly upon the conduct of Newton in the matter. Brewster, in the enlarged edition of his "Life of Newton," devotes a great deal of space to a refutation of this view, and succeeds in proving that, though Newton might have shown occasional irritability of temper, the fault rested chiefly with Flamsteed, who " was prone to take an unfavourable view of the motives as well as the conduct of those with whom he differed, and when such impressions were once made upon his mind, it was almost impossible to dislodge them."—" Newton," he proceeds to remark, " was not in good health during the correspondence. The depths of his mind were stirred with the difficulties of the lunar problem. The new views which burst upon him in its solution could be tested only by observation; and they who have felt the impatience of spirit when a speculation waits for the verdict of an experiment or a fact, or who have started from their midnight couch to submit a happy idea to the ordeal of observation, will understand the sensitiveness of Newton when he waited whole weeks for the precious numbers which the Observatory of Greenwich only could supply." —EDITOR.

of his health, he assures Mr. Pepys "that he is very well,
—that he fears he is under some small degree of melancholy, but that there is no reason to suspect that it hath
at all touched his understanding."

During this period of bodily indisposition, his mind,
though in a state of nervous irritability, and disturbed for
want of rest, was capable of putting forth its highest
powers. At the request of Dr. Wallis he drew up an
example of one of his propositions on the quadrature of
curves in second fluxions. He composed, at the desire of
Dr. Bentley, his profound and beautiful letters on the
existence of the Deity. He was requested by Locke to
reconsider his opinions on the subject of innate ideas.
Dr. Mill engaged him in profound biblical researches; and
we find him grappling with the difficulties of the lunar
theory.

But with all these proofs of a vigorous mind, a diminution of his mental powers has been rashly inferred from
the cessation of his great discoveries, and from his un-
willingness to enter upon new investigations. The facts,
however, here assumed are as incorrect as the inference
which is drawn from them. The ambition of fame is a
youthful passion, which is softened, if not subdued, by
age. Success diminishes its ardour, and early pre-eminence often extinguishes it. Before the middle period of
his life Newton was invested with all the insignia of
immortality; but, endowed with a native humility of
mind, and animated with those hopes which teach us to
form a humble estimate of human greatness, he was
satisfied with the laurels which he had won, and he sought
only to perfect and complete his labours. His mind was
principally bent on the improvement of the Principia;
but he occasionally diverged into new fields of scientific
research—he created his fine theory of astronomical
refractions—he made great improvements on the lunar

theory—he solved problems of great difficulty which had
been proposed to try his strength—he made valuable
additions to his " Opticks "—he continued his chemical
experiments—and he devoted much of his time to profound
inquiries in chronology and in theological literature.

The powers of his mind were therefore in full requisition;
and, when we consider that he was called to the discharge
of high official functions which forced him into public life,
and compelled him to direct his genius into new channels,
we can scarcely be surprised that he ceased to produce
any very original works on abstract science. In the
direction of the affairs of the Mint, and of the Royal
Society, to which we shall now follow him, he found ample
occupation for his time; while the leisure of his declining
years was devoted to those exalted studies in which philosophy yields to the supremacy of faith, and hope admini-
sters to the aspirations of genius.

CHAPTER XIV.

*No Mark of National Gratitude yet conferred upon Newton—
Friendship between him and Charles Montague, afterwards
Earl of Halifax—Mr. Montague appointed Chancellor of
the Exchequer in 1694—He resolves upon a re-Coinage—
Nominates Newton Warden of the Mint in 1695—Newton
appointed Master of the Mint in 1699—Notice of the Earl
of Halifax—Newton elected Associate of the Academy of
Sciences in 1699—Member for Cambridge in 1701—and
President of the Royal Society in 1703—Queen Anne con-
fers upon him the Honour of Knighthood in 1705—Second
Edition of the Principia, edited by Cotes—His Conduct
respecting Mr. Ditton's Method of finding the Longitude.*

HITHERTO we have viewed Newton chiefly as a philosopher leading a life of seclusion within the walls of a
college, and either engaged in the duties of his professorship, or ardently occupied in mathematical and scientific
inquiries. He had now reached the fifty-third year of his
age, and while those of his own standing at the university
had been receiving high appointments in the church, or
lucrative offices in the state, he still remained without any
mark of the respect or gratitude of his country. Though
his friends had exerted themselves to procure him some
permanent appointment, they had failed in the attempt.
An event, however, now occurred which relieved him
from his labours at Cambridge, and placed him in a situation of affluence and honour.

Among his friends at Cambridge Newton had the good
fortune to number Charles Montague, grandson of Henry,

first Earl of Manchester, a young man of high promise,
and every way worthy of his friendship. Though devoted
to literary rather than scientific pursuits, and twenty
years younger than Newton, he cherished for the philosopher all the veneration of a disciple, and his affection
for him gathered new strength whilst himself rising to
the highest honours and offices of the state. In the year
1684 we find him co-operating with Newton in the establishment of a Philosophical Society at Cambridge; but
though both of them had made personal application to
different individuals to induce them to become members,
yet the plan failed, from the want, as Newton expresses
it, of persons willing to try experiments.

Mr. Montague sat along with Newton in the Convention
Parliament, and such were the powers which he displayed
in that assembly as a public speaker, that he was made
a commissioner of the treasury, and soon afterwards a
Privy Councillor. In these situations his talents and
knowledge of business were highly conspicuous, and in
1694 he was appointed Chancellor of the Exchequer.
The current coin of the nation having been adulterated
and debased, one of his earliest designs was to re-coin it,
and restore it to its original value. This scheme, however, like all measures of reform, met with great opposition.
It was characterized as a wild project unsuitable to a
period of war, as highly injurious to the interests of com-
merce, and as likely to sap the foundation of the government. But he had studied the subject too deeply, and
had intrenched himself behind opinions too impartial and
too well founded, to be driven from a measure which the
best interests of his country seemed to require.

The persons whom Mr. Montague had consulted about
the re-coinage were Newton, Locke, and Halley; and
when Mr. Overton, the warden of the mint, was appointed
a commissioner of customs, he embraced the opportunity

which was thus offered of serving his friend and his country by recommending Newton to the important office which thus fell vacant. The appointment was notified in the following letter addressed to him at Cambridge :—

"SIR, *London, 19th of March,* 1695.

"I AM very glad that at last I can give you a good proof of my friendship, and the esteem the king has of your merits. Mr. Overton, the Warden of the Mint, is made one of the Commissioners of the Customs, and the king has promised me to make Mr. Newton Warden of the Mint. The office is the most proper for you. 'Tis the chief office in the Mint. 'Tis worth five or six hundred pounds per annum, and has not too much business to require more attendance than you may spare. I desire you will come up as soon as you can; I will take care of your warrant in the meantime. Pray give my humble services to John Lawton.* I am sorry I have not been able to assist him hitherto, but I hope he will be provided for ere long, and tell him that the session is near ending, and I expect to have his company when I am able to enjoy it. Let me see you as soon as you come to town, that I may carry you to kiss the king's hand. I believe you may have a lodging near me.—I am, &c.

"CHARLES MONTAGUE."

In this new situation the mathematical and chemical knowledge of our author was of great service to the nation, and he became eminently useful in carrying on the re-coinage, which was completed in the short space of two

---

* Mr Lawton or Laughton, was a great personal friend of Sir Isaac Newton and Charles Montague. He was afterwards Librarian and Chaplain of Trinity College. He subsequently became Canon of Worcester and Lichfield, and gave to the Library of Trinity College a valuable collection of books.

years. In the year 1699, he was promoted to the mastership of the Mint,—an office which was worth twelve or
fifteen hundred pounds per annum, and which he held
during the remainder of his life. In this situation he
wrote an official Report on the Coinage, which has been
published; and he drew up a Table of Assays of Foreign
Coins, which is printed at the end of Dr. Arbuthnot's
Tables of Ancient Coins, Weights, and Measures, which
appeared in 1727.

While Newton held the inferior office of warden of the
Mint, he retained the Lucasian chair; but upon his promotion in 1699, he appointed Mr. Whiston to be his deputy
at Cambridge, with " the full profits of the place;" and
when he resigned the professorship on the 10th of December,
1701, he succeeded in getting him elected as his successor.

The appointment of Newton to the mastership of the
Mint must have been peculiarly gratifying to the Royal
Society; and it was probably from a feeling of gratitude
to Mr. Montague, as much as from a regard for his talents,
that this able and accomplished statesman was elected
president of that learned body on the 30th of November,
1695. This office he held for three years, and on the 30th
of January, 1697, Newton had the satisfaction of address-
ing to him his solution of the celebrated problems proposed
by John Bernouilli.

Mr. Montague was created Earl of Halifax in the
year 1700. After the death of his first wife he conceived a strong attachment for Mrs. Catherine Barton,
the widow of Colonel Barton, and the niece of Newton.
This lady was young, gay, and beautiful; and, though she
did not escape the censures of her contemporaries, she was
regarded by those who knew her as a woman of strict
honour and virtue. We are not acquainted with the causes
which prevented her union with the Earl of Halifax; but
so great was the esteem and affection which he bore her,

that in the will in which he left £100 to Newton, he bequeathed to his niece a very large portion of his fortune. This distinguished statesman died in 1715, in the fifty-fourth year of his age. Himself a poet and an elegant writer, he was the liberal patron of genius, and he numbered among his intimate friends Congreve, Halley, Prior, Tickell, Steele, and Pope. His conduct to Newton will be for ever remembered in the annals of science. The sages of every nation and of every age will pronounce with affection the name of Charles Montague; and the persecuted science of England will continue to deplore that he was the first and the last English minister who honoured genius by his friendship and rewarded it by his patronage.

The elevation of Newton to the highest offices in the Mint was followed by other marks of honour. The Royal Academy of Sciences at Paris, having been empowered by a New Charter, granted in 1699, to admit a very small number of foreign associates, Newton was elected a member of that distinguished body. In the year 1701, on the assembling of a new Parliament, he was re-elected one of the members for the University of Cambridge.* In 1703 he was chosen President of the Royal Society of London, and he was annually re-elected to this office during the remaining twenty-five years of his life. On the 16th of April, 1705, when Queen Anne was living at the royal residence of Newmarket, she went with Prince George of Denmark and the rest of the court to visit the university of Cambridge. After the meeting of the Regia Comitia, her majesty held a court at Trinity Lodge, the residence of Dr. Bentley, then master of Trinity, where

* The candidates in 1701 were as follows:—
Mr. Henry Boyle, afterwards Lord Carleton, - 180 } Both of Trinity
Mr. Newton, - - - - - - - 161 } College.
Mr. Hammond, - - - - - - 64

the honour of knighthood was conferred upon Mr. Newton,
Mr. John Ellis, the vice-chancellor, and Mr. James Montague, the university counsel.*

On the dissolution of the Parliament, which took place
in 1705, Sir Isaac was again a candidate for the representation of the university; but notwithstanding the recent
expression of the royal favour, he lost his election by a
very great majority.† This singular result was perhaps
owing to the loss of that personal influence which his
residence in the university could not fail to command;
though it is more probable that the ministry preferred
the candidates of a more obsequious character, and that
the electors looked for advantages which Sir Isaac Newton
was not able to obtain for them.

Although the first edition of the Principia had been for
some time sold off, and the copies of it had become extremely rare, yet Sir Isaac's attention was so much occupied
with his professional avocations, that he could not find
leisure for preparing a new edition. Dr. Bentley, who
had repeatedly urged him to this task, at last succeeded,
by engaging Robert Cotes, Plumian Professor of Astronomy at Cambridge, to superintend its publication at the
university press. In June, 1709, Sir Isaac committed
this important trust to his young friend; and about the
middle of July he promised to send him in the course of a
fortnight, his own revised copy of the work. Business,
however, seems to have intervened, and Mr. Cotes was

* The banquet which was on this occasion given in the College hall to
the Royal visitor seems to have cost about £1000, and the University was
obliged to borrow £500 to defray the expense of it.—Monk's *Life of
Bentley*, pp. 143, 144.

† The candidates in 1705 were as follows :—

The Hon. Arthur Annesley, - - -	182
Hon. Dixie Windsor, - - -	170
Mr. Godolphin, - - - -	162
Sir Isaac Newton, - - - -	117

obliged to remind Sir Isaac of his promise, which he did
in the following letter:—

"SIR,   *Cambridge, Aug. 18th,* 1709.

"THE earnest desire I have to see a new edition of your
Principia, makes me somewhat impatient till we receive
your copy of it, which you were pleased to promise me
about the middle of last month you would send down in
about a fortnight's time.   I hope you will pardon me for
this uneasiness, from which I cannot free myself, and for
giving you this trouble to let you know it.   I have been
so much obliged to yourself and your book, that (I desire
you to believe me) I think myself bound in gratitude to
take all the care I possibly can that it shall be correct.—
Your obliged servant,

                            "ROGER COTES.

"*For Sir Isaac Newton, at his house
    in Jermyn Street, near St. James'
    Church, Westminster.*"

This was the first letter of that celebrated correspondence, consisting of nearly three hundred letters, in which
Sir Isaac and Mr. Cotes discussed the various improvements which were thought necessary in a new edition of
the Principia.   This valuable collection of letters is preserved in the Library of Trinity College; and we cannot
refrain from repeating the wish expressed by Dr. Monk,
"that one of the many accomplished Newtonians who are
resident in that society would favour the world by publishing the whole collection."

When the work was at last printed, Mr. Cotes expressed
a wish that Dr. Bentley should write the preface to it, but
it was the opinion both of Sir Isaac and Dr. Bentley,
that the preface should come from the pen of Mr. Cotes
himself.   This he accordingly undertook; but previous to its

execution he addressed the following letter to Dr. Bentley, in order to learn from Sir Isaac the particular view with which it should be written.

"SIR,   *March* 10*th*, 1712–13.

"I RECEIVED what you wrote to me in Sir Isaac's letter. I will set about the index in a day or two. As for the preface, I should be glad to know from Sir Isaac with what view he thinks proper to have it written. You know the book has been received abroad with some disadvantage, and the cause of it may be easily guessed at. The *Commercium Epistolicum*, lately published by order of the Royal Society, gives such indubitable proofs of Mr. Leibnitz's want of candour, that I shall not scruple in the least to speak out the full truth of the matter, if it be thought convenient. There are some pieces of his looking this way, which deserve a censure, as his *Tentamen de Motuum cœlestium Causis.* If Sir Isaac is willing that something of this nature may be done, I should be glad, if, whilst I am making the index, he would consider of it, and put down a few notes of what he thinks most material to be insisted on. This I say upon supposition that I write the preface myself. But I think it will be much more advisable that you, or he, or both of you, should write it whilst you are in town. You may depend upon it I will own it, and defend it as well as I can, if hereafter there be occasion.—I am, Sir," &c.

We are not acquainted with the instructions which were given to Mr. Cotes in consequence of this application; but it appears from the preface itself, which contains a long and able summary of the Newtonian philosophy, that Sir Isaac had prohibited any personal reference to the conduct of Leibnitz.

The general preface is dated May 12th, 1713, and in a

subsidiary preface of only a few lines, dated March 28th, 1713, Sir Isaac mentions the leading alterations which had been made in this edition. The determination of the forces by which bodies may revolve in given orbits was simplified and enlarged. The theory of the resistance of fluids was more accurately investigated, and confirmed by new experiments. The theory of the moon, and the precession of the equinoxes, were more fully deduced from their principles; and the theory of comets was confirmed by several examples of their orbits more accurately computed.

In the year 1714* several captains and owners of merchant vessels petitioned the House of Commons to consider the propriety of bringing in a Bill to reward inventions for promoting the discovery of the longitude at sea. A Committee was appointed to investigate the subject, and Mr. Ditton and Mr. Whiston, having thought of a new method of finding the longitude, submitted it to the Committee. Four members of the Royal Society, viz., Sir Isaac Newton, Dr. Halley, Mr. Cotes, and Dr. Clarke, were examined on the subject, along with Mr. Ditton and Mr. Whiston. The last three of these philosophers stated their opinions verbally. Mr. Cotes considered the proposed scheme as correct in theory and on shore, and both he and Dr. Halley were of opinion that expensive experiments would be requisite. Newton, when called upon for

---

* It was about forty years before this that a French charlatan (Le Sieur de St. Pierre) had procured from Charles II. a commission for examining a scheme for the discovery of the longitude. In 1675, Flamsteed, being then a member of the commission, wrote a letter to the commissioners explanatory of his views, which, being shown to the King, led to the establishment of the Royal Observatory at Greenwich, to which Flamsteed himself was made the "astronomical observator," with the object of "rectifying the tables of the motions of the heavens, and the places of the fixed stars, so as to find out the so much desired longitude of places, for the perfecting the art of navigation."—EDITOR.

his opinion, read the following memorandum, which deserves to be recorded :—

"For determining the longitude at sea there have been several projects, true in theory, but difficult to execute.

"1. One is by a watch to keep time exactly; but, by reason of the motion of the ship, the variation of heat and cold, wet and dry, and the difference of gravity in different latitudes, such a watch hath not yet been made.

"2. Another is by the eclipses of Jupiter's satellites; but, by reason of the length of telescopes requisite to observe them, and the motion of a ship at sea, those eclipses cannot yet be there observed.

"3. A third is by the place of the moon; but her theory is not yet exact enough for that purpose; it is exact enough to determine the longitude within two or three degrees, but not within a degree.

"4. A fourth is Mr. Ditton's project, and this is rather for keeping an account of the longitude at sea than for finding it, if at any time it should be lost, as it may easily be in cloudy weather. How far this is practicable, and with what charge, they that are skilled in sea affairs are best able to judge. In sailing by this method, whenever they are to pass over very deep seas, they must sail due east or west, without varying their latitude; and if their way over such a sea doth not lie due east or west, they must first sail into the latitude of the next place to which they are going beyond it, and then keep due east or west, till they come at that place.

"In the three first ways there must be a watch regulated by a spring, and rectified every visible sunrise and sunset, to tell the hour of the day or night. In the fourth way such a watch is not necessary. In the first way there must be two watches, this and the other above mentioned.

"In any of the three first ways, it may be of some service to find the longitude within a degree, and of much more

service to find it within forty minutes, or half a degree
if it may be, and the success may deserve rewards accordingly.

"In the fourth way, it is easier to enable seamen to know
their distance and bearing from the shore, forty, sixty, or
eighty miles off, than to cross the seas; and some part of
the reward may be given, when the first is performed on
the coast of Great Britain, for the safety of ships coming
home; and the rest, when seamen shall be enabled to sail
to an assigned remote harbour without losing their longitude if it may be."

The committee brought up their report on the 11th of
June, stating "that it is the opinion of this committee
that a reward be settled by Parliament upon such person
or persons as shall discover a more certain and practicable
method of ascertaining the longitude than any yet in
practice; and the said reward be proportioned to the degree
of exactness to which the said method shall reach." This
resolution was unanimously adopted by the House. A
bill embodying it was accordingly framed from it, and
after passing the House of Commons on the 3rd of July,
was agreed to by the Lords on the 8th of the same month.[*]

In giving an account of this transaction,[†] Mr. Whiston
states, that nobody understood Sir Isaac's paper, and that
after sitting down he obstinately kept silence, though he
was much pressed to explain himself more distinctly. At
last seeing that the scheme was likely to be rejected,
Whiston ventured to say that Sir Isaac did not wish to
explain more through fear of compromising himself, but
that he really approved of the plan, knowing the usefulness of the present method near the shores, (which are the
places of greatest danger.) Sir Isaac, he goes on to say,

* Journals of the House of Commons, vol. xvii., pp. 641, 671, 677, and
716.

† Whiston's "Longitude discovered." Lond., 1738.

stood up and said that "he thought this bill ought to pass, because of the present method's usefulness near the shores." This is the part of Newton's conduct which M. Biot has described as puerile, and "tending to confirm the fact of the aberration of his intellect in 1693." Before we can admit such a censure, we must be satisfied with the correctness of Whiston's statement. Newton's paper is perfectly intelligible, and we may easily understand how he might have approved of Mr. Ditton's plan as ingenious and practicable under particular circumstances, though he did not think it of that paramount importance which would have authorized the House of Commons to distinguish it by a parliamentary reward. The conflict between public duty and a disposition to promote the interests of Mr. Whiston and Mr. Ditton, was no doubt the cause of that embarrassment of manner which the former of these mathematicians has so unkindly brought before the public. The effect of Newton's opinion upon the committee must have been highly gratifying to himself and his friends; and when he simply paused in repeating orally what he had so distinctly read from his paper, he little thought that a future biographer would ascribe an interval of silence to "puerility of conduct"—to "an inexplicable timidity of mind," and to "the consequences of a previous mental aberration."

CHAPTER XV.

On the accession of George I. to the British throne in
1714, Sir Isaac Newton became an object of interest at
court, his friend and patron, Lord Halifax, being now
First Lord of the Treasury. His own high situation
under government,—his splendid reputation,—his spotless
character,—and, above all, his unaffected piety, attracted
the attention of the Princess of Wales, afterwards Queen
Consort to George II. This lady, who possessed a highly
cultivated mind, derived the greatest pleasure from conversing with Newton and corresponding with Leibnitz.
In all her difficulties, she received from Sir Isaac that
information and assistance which she had elsewhere sought
in vain; and she was often heard to declare in public, that
she thought herself fortunate in living at a time which

enabled her to enjoy the conversation of so great a genius.
But while Newton was thus esteemed by the house of
Hanover, Leibnitz, his great rival, endeavoured to weaken
and undermine his influence.   In his correspondence with
the Princess, he represented the Newtonian philosophy
not only as physically false, but as injurious to the interests
of religion.   He asserted that natural religion was rapidly
declining in England, and he supported this position by
referring to the works of Locke, and to the beautiful and
pious sentiments contained in the twenty-eighth Query at
the end of the Optics.   He represented the principles of
these great men as precisely the same with those of the
materialists, and thus endeavoured to degrade the character
of English philosophers.

These attacks of Leibnitz became subjects of conversation at court; and when they reached the ear of the King,
his majesty expressed a wish that Sir Isaac Newton would
draw up a reply.   He accordingly entered the lists on the
mathematical part of the controversy, and left the philosophical part of it to Dr. Clarke, who was a full match
for the German philosopher.   The correspondence which
thus took place was carefully perused by the Princess, and
from the estimation in which Sir Isaac continued to be
held, we may infer that the views of the English philosopher were not very remote from her own.

When Sir Isaac was one day conversing with her Royal
Highness on some points of ancient history, he was led to
mention to her, and to explain, a new system of chronology
which he composed during his residence at Cambridge,
where he was in the habit, as he himself expresses it, " of
refreshing himself with history and chronology when he
was weary of other studies."   The Princess was so much
pleased with the ingenuity of his system, that she subsequently, in the year 1718, sent a message by the Abbé
Conti to Sir Isaac, desiring him to speak with her; and,

on this occasion, she requested a copy of the interesting
work which contained his system of chronology. Sir Isaac
informed her that it existed only in separate papers, which
were not only in a state of confusion, but contained a very
imperfect view of the subject; and he promised in a few
days to draw up an abstract of it for her own private use,
and on the condition that it should not be communicated
to any other person. Some time after the Princess received
the manuscript, she requested that the Abbé Conti might
be permitted to have a copy of it. Sir Isaac granted this
request, and the Abbé was distinctly informed that the
manuscript was given to him at the request of the Princess,
and with Sir Isaac's leave, and that he was to keep it a
secret.* The manuscript, which was thus rashly put into
the hands of a foreigner, was entitled "A Short Chronicle
from the First Memory of Things in Europe to the Conquest of Persia by Alexander the Great." It consists of
about twenty-four quarto printed pages,† with an introduction of four pages, in which Sir Isaac states that he
"does not pretend to be exact to a year, that there may
be errors of five or ten years, and sometimes twenty, but
not much above."

The Abbé Conti kept his promise of secrecy during his
residence in England, but he no sooner reached Paris,
than he communicated it to M. Freret, a learned antiquarian, who not only translated it, but drew up observations
upon it for the purpose of refuting some of its principal
results. Sir Isaac was unacquainted with this transaction
till he was informed of it by the French bookseller,

* This anecdote concerning the chronological manuscript is not correctly
given in the Biographia Britannica, and in some of the other lives of
Newton. I have followed implicitly Newton's own account of it in the
Phil. Trans., 1725, vol. xxxiii., No. 389, p. 315.

† M. Biot has supposed that this abstract was an imperfect edition of
Newton's work on Chronology.

M. Cavelier, who requested his leave to publish it, and charged one of his friends in London to procure Sir Isaac's answer, which was as follows:—

"I remember that I wrote a Chronological Index for a particular friend, on condition that it should not be communicated. As I have not seen the manuscript which you have under my name, I know not whether it be the same. That which I wrote was not at all done with design to publish it. I intend not to meddle with that which hath been given you under my name, nor to give any consent to the publishing of it.—I am your very humble servant,   "ISAAC NEWTON.

"*London, May 27th,* 1725, O. S."

Before this letter was written, viz., on the 21st of May, the bookseller had received the royal privilege for printing the work; and when it was completed, he sent a copy as a present to Sir Isaac, who received it on the 11th of November, 1725. It was entitled, *Abrégé de Chronologie de M. Le Chevalier Newton, fait par lui-même, et traduit sur le manuscrit Anglais,* and was accompanied with observations by M. Freret,* the object of which was to refute the leading points of the system.† An advertisement was prefixed to it, in which the bookseller defends himself for printing it without the author's leave, on the ground that he had written three letters to obtain permission, and had declared that he would take Sir Isaac's silence for consent. When Sir Isaac received this work,

* Father Souciet was supposed by Halley and others to have been the author of these observations, but there is no doubt that they were written by M. Freret.

† It is stated in the *Biogr. Britannica,* Art. *Newton,* that the copy of the French translation was not accompanied by the refutation. Though the reverse of this is not distinctly stated by Sir Isaac himself, yet it may be inferred from his observations.

he drew up a paper entitled, *Remarks on the Observations made on a Chronological Index of Sir Isaac Newton, translated into French by the Observator, and published at Paris,* which was printed in the *Philosophical Transactions* for 1725.* In this paper Sir Isaac gives a history of the transaction,—charges the Abbé Conti with a breach of promise, and blames the publisher for having asked his leave to print the translation without sending him a copy for his perusal,—without acquainting him with the name of the translator,—and without announcing his intention of printing along with it a refutation of the original. The observations made by the translator against the conclusions deduced by the author were founded on an imperfect knowledge of Sir Isaac's system; and they are so specious, that Halley himself confesses that he was at first prejudiced in favour of the Observations, taking the calculations for granted, and not having seen Sir Isaac's work.

To all the observations of M. Freret Sir Isaac returned a triumphant answer. This presumptuous antiquarian had ventured to state at the end of his observations, " that he believed he had stated enough concerning the epochs of the Argonauts, and the length of generations, to make people cautious about the rest; for these are the two foundations of all this new system of chronology." He founds his arguments against the epochs of the Argonauts, as fixed by our author, on the supposition that Sir Isaac places the vernal equinox at the time of the Argonautic expedition *in the middle of the sign of Aries,* whereas Sir Isaac places it *in the middle of the constellation,*—a point corresponding with the middle of the back of Aries, or 8° from the first star of Aries. This position of the colure is assigned on the authority of Eudoxus, as given by

* Vol. xxxiii., No. 389, p. 315.

Hipparchus, who says that the colure passed over the back of Aries. Setting out with this mistake, M. Freret concludes that the Argonautic expedition took place 532 years earlier than Sir Isaac made it. His second objection to the new system relates to the length of generations, which he says is made only eighteen or twenty years. Sir Isaac, on the contrary, reckons a generation at thirty-three years, or three generations at one hundred; and it was the lengths of the reigns of kings that he made eighteen or twenty years. This deduction he founds on the reigns of sixty-four French kings. Now, the ancient Greeks and Egyptians reckoned the length of a reign equal to that of a generation; and it was by correcting this mistake, and adopting a measure founded on fact, that Sir Isaac placed the Argonautic expedition forty-four years after the death of Solomon, and fixed some of the other points of his system.

This answer of Sir Isaac's to the objections of Freret called into the field a fresh antagonist, Father Souciet, who published five Dissertations on the new chronology. These Dissertations were written in a tone highly reprehensible; and the friends of Sir Isaac being apprehensive that the manner in which his system was attacked would affect him more than the arguments themselves, prevailed upon a friend to draw up an abstract of Souciet's objections, stripped of the "extraordinary ornaments with which they were clothed." The perusal of these objections had no other effect upon him than to convince him of the ignorance of their author; and he was induced to read the entire work, which produced no change in his opinion.

In consequence of these discussions, Sir Isaac was prevailed upon to prepare his larger work for the press. He had nearly completed it at the time of his death, and it was published in 1728 under the title of *The Chronology* *of Ancient Kingdoms amended, to which is prefixed a short Chronicle, from the first Memory of Things in Europe to the Conquest of Persia by Alexander the Great.* It was dedicated to the Queen by Mr. Conduitt, and consists of six chapters: 1. On the Chronology of the Greeks;[*] 2. Of the Empire of Egypt; 3. Of the Assyrian Empire; 4. Of the two contemporary Empires of the Babylonians and Medes; 5. A description of the Temple of Solomon; 6. Of the Empire of the Persians. The sixth chapter was not copied out with the other five, which makes it doubtful whether or not it was intended for publication; but as it was found among his papers, and appeared to be a continuation of the same work, it was thought right to add it to the other five chapters.[†]

After the death of Newton, Dr. Halley, who had not yet seen the larger work, felt himself called upon, both as Astronomer-Royal,[‡] and as the friend of the author, to reply to the first and last Dissertations of Father Souciet, which were chiefly astronomical; and in two papers, printed in the Philosophical Transactions for 1727,[‖] he has done this in a most convincing and learned argument.

Among the supporters of the views of Newton, we may enumerate Dr. Reid, Nauze, and some other writers; and

---

[*] According to Whiston, Sir Isaac wrote out eighteen copies of this chapter with his own hand, differing little from one another.—*Whiston's Life*, p. 89.

[†] This work forms the first article in the fifth volume of Dr. Horsley's edition of Newton's works. The next article in the volume is entitled, "A Short Chronicle from a MS., the property of the Reverend Dr. Ekins, Dean of Carlisle," which is nothing more than the abstract of the Chronology already printed in the same volume. We cannot even conjecture the reasons for publishing it, especially as it is less perfect than the abstract, two or three dates being wanting.

[‡] He had succeeded to that office early in 1720, shortly after the death of Flamsteed in December, 1719, and held it until his own death in January, 1742.—Editor.

[‖] See vol. xxxiv., p. 205, and vol. xxxv., p. 296.

among his opponents, M. Freret, who left behind him a
posthumous work on the subject, M. Fourmond, Mr. A.
Bedford, Dr. Shuckford, Dr. Middleton, Whiston, and the
late M. Delambre. The object of M. Fourmond is to
show the uncertainty of the astronomical argument, arising
on the one hand from the vague account of the ancient
sphere as given by Hipparchus; and, on the other, from
the extreme rudeness of ancient astronomical observations. Delambre has taken a similar view of the subject:
he regards the observations of ancient astronomers as too
incorrect to form the basis of a system of chronology; and
he maintains that, if we admit the accuracy of the details
in the sphere of Eudoxus, and suppose them all to belong
to the same epoch. all the stars which it contains ought at
that epoch to be found in the place where they are marked,
and we might thence verify the accuracy, and ascertain
the state, of the observations. It follows, however, from
such an examination, that the sphere would indicate almost as many different epochs as it contains stars. Some
of them had not, in the time of Eudoxus, even arrived at
the position which had been for a long time attributed to
them, and will not indeed reach it for three hundred years
to come, and on this account he considers it impossible to
deduce any chronological conclusions from such a rude
mass of errors.

But, however well founded these observations may be,
we agree in opinion with M. Daunou,* "that they are
not sufficient to establish a new system, and we must
regard the system of Newton as a great fact in the history
of chronological science, and as confirming the observation
of Varro, that the stage of history does not commence
till the first Olympiad."

* See an excellent view of this chronological controversy in an able
note by M. Daunou, attached to Biot's Life of Newton in the *Biogr.
Universelle,* tom. xxxi., p. 180.

Among the chronological writings of Sir Isaac Newton, we must enumerate his *Letter to a person of distinction who had desired his opinion of the learned Bishop Lloyd's Hypothesis concerning the Form of the most Ancient Year.* This hypothesis was sent by the Bishop of Worcester to Dr. Prideaux. Sir Isaac remarks, that it is filled with many excellent observations on the ancient year; but he does not "find it proved that any ancient nation used a year of twelve months and three hundred and sixty days, without correcting it from time to time by the luminaries, to make the months keep to the course of the moon, and the year to the course of the sun, and returns of the seasons and fruits of the earth." After examining the years of all the nations of antiquity, he concludes, that "no other years are to be met with among the ancients but such as were either luni-solar, or solar or lunar, or the calendars of these years." "A practical year," he adds, "of three hundred and sixty days, is none of these. The beginning of such a year would have run round the four seasons in seventy years, and such a notable revolution would have been mentioned in history, and is not to be asserted without proving it." *

* This letter is published without any date in the *Gentleman's Magazine* for 1755, vol. xxv., p. 8. It bears internal evidence of being genuine.

CHAPTER XVI.

THE history of the theological studies of Sir Isaac Newton will ever be regarded as one of the most interesting portions of his life. That the greatest philosopher of which any age can boast was a sincere and humble believer in the leading doctrines of our religion, and lived conformably to its precepts, has been justly regarded as a proud triumph of the Christian faith. Had he exhibited only an outward respect for the forms and duties of religion, or left merely in his last words an acknowledgment of his belief, his piety might have been regarded as a prudent submission to popular feeling, or as a proof of the decay or extinction of his transcendent powers. But he had been a searcher of the Scriptures from his youth, and he found it no abrupt transition to pass from the study of the material universe to an investigation of the profoundest truths, and the most obscure predictions, of Holy Writ.

As the religious habits of Sir Isaac Newton could not be ascribed to an ambition of popularity, to the influence of weak health, or to the force of professional impulse, it became necessary for the apostles of infidelity to refer it to some extraordinary cause. His supposed insanity was, therefore, eagerly seized upon by some as affording a plausible origin for his religious principles; while others, without any view of supporting the cause of scepticism, ascribed his theological researches to the habits of the age in which he lived, and to a desire of promoting political liberty, by turning against the abettors of despotism those powerful weapons which the Scriptures supplied. The anxiety evinced by M. de La Place to refer his religious writings to a late period of his life, seems to have been felt also by M. Biot, who has gone so far as to fix the very date of one of his most important works, and to associate his religious tendencies with the effects of what he calls "the fatal epoch of 1693."

"From the nature of the subject," says he, "and from certain indications which Newton seems to give at the beginning of his dissertation,* we may conjecture with probability that he composed it at the time when the errors of Whiston, and a work of Dr. Clarke on the same subject, drew upon them the attacks of all the theologians of England, which would place the date between the years 1712 and 1719.† It would then be truly a prodigy to remark, that a man of from seventy-two to seventy-five years of age was able to compose, *rapidly*, as he leads us to believe, so extensive a piece of sacred criticism, of literary history, and even of bibliography, where an erudition the most vast, the most varied, and the most ready, always supports an argument well arranged and powerfully combined . . . At this epoch of the life of Newton the reading of religious books had become one of his most habitual occupations; and after he had performed

* His *Historical Account of two notable Corruptions of Scripture.* 50 pp. Quarto.

† The difference of these dates, as well as the narrower limits set down for Newton's age, shew what guess-work this is.—EDITOR.

the duties of his office, they formed, along with the
conversation of his friends, his principal amusement.   He
had then almost ceased to care for the sciences; and, as we
have already remarked, since the fatal epoch of 1693, he
gave to the world only three really new scientific
productions."

Notwithstanding the prodigy which it involves, M. Biot
has adopted 1712-1719 as the date of this critical dissertation;—it is regarded as the composition of a man of
seventy-two or seventy-five;—the reading of religious
works is stated to have *become* one of his most habitual
occupations, and such reading is said to have been one of
his principal amusements; and all this is associated with
"the fatal epoch of 1693," as if his illness at that time
had been the cause of his abandoning science and betaking
himself to theology.   Carrying on the same views, M.
Biot asks, in reference to Sir Isaac's work on Prophecy,
"How a mind of the character and force of Newton's, so
habituated to the severity of mathematical considerations,
so exercised in the observation of real phenomena, and
so well aware of the conditions by which truth is to be
discovered, could put together such a number of conjec-
ures without noticing the extreme improbability of his
interpretations from the infinite number of arbitrary pos-
tulates on which he has founded them?"   We would
apply the same question to the reasoning by which M.
Biot fixes the date of the critical dissertation; and we
would ask how so eminent a philosopher could hazard
such frivolous conjectures upon a subject on which he had
not a single fact to guide his inquiries.   The obvious
tendency, though not the design, of the conclusion at
which he arrives, is injurious to the memory of Newton,
as well as to the interests of religion; and these considerations might have checked the temerity of speculation,
even if it had been founded on better data.   The New-

tonian interpretation of the Prophecies, and especially
that part which M. Biot characterizes as unhappily stamped
with the spirit of prejudice, has been adopted by men of
the soundest and most unprejudiced minds; and, in addition to the moral and historical evidence by which it is
supported, it may yet be exhibited in all the fulness of
demonstration. But the speculation of Biot respecting
the date of Newton's theological works was never maintained by any other person than himself, and is capable
of being disproved by the most incontrovertible evidence.

We have already seen in the extract from Mr. Pryme's
manuscript, that previous to 1692, when a shade is supposed to have passed over his gifted mind, Newton was
well known by the appellation of an " excellent divine," a
character which could not have been acquired without the
devotion of many years to theological researches; but,
important as this argument would have been, we are fortunately not left to so general a defence. The correspondence of Newton with Locke, recently published by Lord
King, places it beyond a doubt that he had begun his
researches respecting the prophecies before the year 1691,
before the forty-ninth year of his age, and before the
" fatal epoch of 1693." The following letter shows that
he had previously discussed this subject with his friend :—

" SIR,   *Cambridge, February 7th,* 1690–1.
" I AM sorry your journey proved to so little purpose,
though it delivered you from the trouble of the company
the day after. You have obliged me by mentioning me
to my friends at London, and I must thank both you and
my Lady Masham for your civilities at Oates, and for not
thinking that I made a long stay there. I hope we shall
meet again in due time, and then I should be glad to have
your judgment upon some of my mystical fancies. The
Son of Man (Dan. vii.), I take to be the same with the

Word of God upon the White Horse in Heaven (Apoc.
xix.), for both are to rule the nations with a rod of iron;
but whence are you certain that the Ancient of Days is
Christ?   Does Christ anywhere sit upon the throne?   If
Sir Francis Masham be at Oates, present, I pray, my
service to him, with his lady, Mrs. Cudworth, and Mrs.
Masham.   Dr. Covel is not in Cambridge.   I am, your
affectionate and humble servant,   "Is. Newton."

'Know you the meaning of Dan. x. 21?—'*There is
none that holdeth with me in these things but Mich. your
Prince.'*"

Having thus determined the date of those investigations
which constitute his *Observations on the Prophecies of Holy
Writ,* particularly the prophecies of Daniel and the
Apocalypse, we shall proceed to fix the latest date of
his *Historical Account of two notable Corruptions of
Scripture, in a Letter to a Friend.*

This work seems to have been a *very early* production
of our author.   It was written in the form of a letter to
Mr. Locke, and at that time Sir Isaac seems to have been
anxious for its publication.   Afraid, however, of being
again led into a controversy, and dreading the intolerance
to which he might be exposed, he requested Mr. Locke,
who was at that time meditating a voyage to Holland, to
get it translated into French, and published on the Continent.   Having abandoned his design of visiting Holland,
Locke transmitted the manuscript, without Newton's name,
to his learned friend M. Le Clerc, in Holland; and it
appears from a letter of Le Clerc's to Locke, that he must
have received it before the 11th of April, 1691.   M. Le
Clerc delayed for a long time to take any steps regarding
its publication; but in a letter dated January 20th, 1692,
he announced to Locke his intention of publishing the
tract in Latin.   When this plan was communicated to

Sir Isaac, he became alarmed at the risk of detection, and resolved to stop the publication of his manuscript. This resolution was intimated to Mr. Locke in a letter from which the following is an extract:—

"Sir,   *Cambridge, Feb.* 16*th,* 1691-2.

"Your former letters came not to my hand, but this I have. I was of opinion my papers had lain still, and am sorry to hear there is news about them. Let me entreat you to stop their translation and impression so soon as you can; for I design to suppress them. If your friend hath been at any pains and charge, I will repay it, and gratify him. . . . . . .
Your most affectionate and humble servant,
        "Is. Newton.

Hence we see that this celebrated treatise, which Biot alleges to have been written between 1712 and 1719, was actually in the hands of Le Clerc in Holland previous to the 11th of April, 1691, and consequently previous to the time of the supposed insanity of its author. Mr. Locke lost no time in obeying the request of his friend. Le Clerc instantly stopped the publication of the letter, and, as he had never learned the name of the author, he deposited the manuscript, which was in the handwriting of Mr. Locke, in the Library of the Remonstrants, where it was afterwards found, and was published at London in 1754, under the title of *Two Letters from Sir Isaac Newton to M. Le Clerc,* a form which had never been given to it by its author. The copy thus published was a very imperfect one, wanting both the beginning * and the end, and erroneous in many places; but Dr. Horsley has published a genuine edition, which has the form of a single

* The editor supplied the beginning down to the thirteenth page where he mentions in a note, that "*thus far is not Sir Isaac's.*"

letter to a friend, and was copied from a manuscript in
Sir Isaac Newton's handwriting, now in the possession of
the Rev. J. Ekins, Rector of Sampford.

Having thus determined, as accurately as possible, the
dates of the principal theological writings of Sir Isaac,
we shall now proceed to give some account of their contents.

The work entitled *Observations upon the Prophecies of
Daniel and the Apocalypse of St. John*, was published in
London in 1733, in one volume, quarto. It is divided
into two parts, the first of which treats of the Prophecies
of Daniel, and the second of the Apocalypse of St. John.
It begins with an account of the different books which
compose the Old Testament; and, as the author considers
Daniel to be the most distinct in the order of time, and
the easiest to be understood, he makes him the key to all
the prophetic books in those matters which relate to the
" last time." He next considers the figurative language
of the prophets, which he regards as taken " from the
analogy between the world natural, and an empire or
kingdom considered as a world politic;" the heavens, and
the things therein, representing thrones and dynasties; the
earth, with the things therein, the inferior people; and
the lowest parts of the earth, the most miserable of the
people. The sun is put for the whole race of kings, the
moon for the body of the common people, and the stars
for subordinate princes and rulers. In the earth, the dry
land and the waters are put for the people of several
nations. Animals and vegetables are also put for the
people of several regions. When a beast or man is put
for a kingdom, his parts and qualities are put for the
analogous parts and qualities of the kingdom; and when
a man is taken in a mystical sense, his qualities are often
signified by his actions, and by the circumstances and

things about him.   In applying these principles he begins
with the vision of the image composed of four different
metals.   This image he considers as representing a body
of four great nations which should reign in succession
over the earth, viz., the people of Babylonia, the Persians,
the Greeks, and the Romans; while the stone cut out
without hands is a new kingdom which should arise after
the four, conquer all those nations, become very great, and
endure to the end of time.

The vision of the four beasts is the prophecy of the
four empires repeated, with several new additions.   The
lion with eagles' wings was the kingdom of Babylon and
Media, which overthrew the Assyrian power.   The beast
like a bear was the Persian empire, and its three ribs were
the kingdoms of Sardis, Babylon, and Egypt.   The third
beast, like a leopard, was the Greek Empire, and its four
heads and four wings were the kingdoms of Cassander,
Lysimachus, Ptolemy, and Seleucus.   The fourth beast,
with its great iron teeth, was the Roman empire, and its
ten horns were the ten kingdoms into which it was broken
in the reign of Theodosius the Great.

In the fifth chapter Sir Isaac treats of the kingdoms
represented by the feet of the image composed of iron and
clay which did not stick to one another, and which were
of different strength.   These were the Gothic tribes
called Ostrogoths, Visigoths, Vandals, Gepidæ, Lombards,
Burgundians, Alans, &c., all of whom had the same
manners and customs, and spoke the same language; and
who, about the year 416 A.C., were all quietly settled in
several kingdoms within the empire, not only by conquest,
but by grants of the Emperor.

In the sixth chapter he treats of the *ten* kingdoms represented by the ten horns of the fourth beast, into which
the Western empire became divided about the time when

Rome was besieged and taken by the Goths. These kingdoms were,—

1. The kingdom of the Vandals and Alans in Spain and Africa.

2. The kingdom of Suevians in Spain.

3. The kingdom of the Visigoths.

4. The kingdom of the Alans in Gaul.

5. The kingdom of the Burgundians.

6. The kingdom of the Franks.

7. The kingdom of the Britons.

8. The kingdom of the Huns.

9. The kingdom of the Lombards.

10. The kingdom of Ravenna.

Some of these kingdoms at length fell, and new ones sprung up; but whatever was their subsequent number, they still retain the name of the ten kings from their first number.

The eleventh horn of Daniel's fourth beast is shown in chapter vii. to be the church of Rome in its triple character of a seer, a prophet, and a king, and its power to change times and laws is copiously illustrated in chapter viii.

In the ninth chapter our author treats of the kingdom represented in Daniel by the ram and he-goat; the ram indicating the kingdom of the Medes and Persians, from the beginning of the four empires, and the he-goat the kingdom of the Greeks to the end of them.

The prophecy of the seventy weeks, which had hitherto been restricted to the first coming of our Saviour, is shown to be a prediction of all the main periods relating to the coming of the Messiah, the times of his birth and death, the time of his rejection by the Jews, the duration of the Jewish war, by which he caused the city and sanctuary to be destroyed, and the time of his second coming.

In the eleventh chapter Sir Isaac treats with great sagacity and acuteness of the time of our Saviour's birth and

passion,*—a subject which had perplexed all preceding
commentators.

After explaining in the twelfth chapter the last prophecy of Daniel, namely, that of the scripture of truth,
which he considers as a commentary on the vision of the
ram and he-goat, he proceeds in the thirteenth chapter of
the prophecy of the king who did according to his will, and
magnified himself above every god, and honoured Mahuz-
zims, and regarded not the desire of women. He shows
that the Greek empire, after the division of the Roman
empire into the Greek and Latin empires, became the king
who, in matters of religion, did according to his will, and
in legislation exalted and magnified himself above every
god.

In the second part of his work, viz., *On the Apocalypse
of St. John*, Sir Isaac treats, 1st: Of the time when the
prophecy was written, which he conceives to have been
during John's exile in Patmos, and before the Epistle to
the Hebrews and the Epistles of Peter were written, which

* It may not be inappropriate here to call attention to two pamphlets
recently published by the Rev. J. Nowland Smith, M.A., of Greenwich,
on Easter-Tide, in which (although that is not the principal subject of
his papers) he discusses the question of the actual time of our Lord's
Crucifixion, and corrects some mistakes which had obtained currency,
and led to the opinion that the date of that great event was A.D. 33,
whereas in all probability it should be A.D. 30. Now the first three
Gospels leave us in no doubt that the Crucifixion took place on a Friday,
and also on the day following the Jewish passover, or Full Moon. Calculation shows that a full moon must have occurred on the night (between
9 and 10 o'clock) of April 6th, in the year 30. This was a paschal full
moon; and this also fell (as Mr. Newland Smith, rightly correcting Mr.
Greswell in this respect, remarks) on a Thursday, the day of the week
preceding that on which our Lord was crucified, and the actual day of
His keeping His last passover. In regard to the bearing of this upon
prophecy, we may remark that, as " the commandment " (of Artaxerxes)
" to restore and to build Jerusalem" appears to have been issued B.C. 454,
from that time to A.D. 30, 483 years, or 69 (" seven and three-score and
two") weeks of years must have elapsed, and the seventieth week of
years have been entered upon, at the time of the Crucifixion.—EDITOR.

in his opinion have a reference to the Apocalypse; 2ndly:
Of the scene of the vision, and the relation which the
Apocalypse has to the book of the law of Moses, and to
the worship of God in the temple; and, 3rdly: Of the
relation which the Apocalypse has to the prophecies of
Daniel, and of the subject of the prophecy itself.

Sir Isaac regards the prophecies of the Old and New
Testament not as given to gratify men's curiosities, by
enabling them to foreknow things, but that, after they
were fulfilled, they might be interpreted by the event, and
afford convincing arguments that the world is governed
by Providence. He considers that there is so much of
this prophecy already fulfilled, as to afford to the diligent
student sufficient instances of God's Providence; and he
adds, that "amongst the interpreters of the last age, there
is scarce one of note who hath not made some discovery
worth knowing, and thence it seems one may gather that
God is about opening these mysteries. The success of
others," he continues, "put me upon considering it, and
if I have done any thing which may be useful to following
writers, I have my design." Such is a brief abstract of
this ingenious work, which is characterised by great
learning, and marked with the sagacity of its distinguished
author. The same qualities of his mind are equally conspicuous in his *Historical Account of Two Notable Corruptions of Scripture*.

This celebrated treatise relates to two texts in the
Epistles of St. John and St. Paul. The first of these is
in 1 John v. 7, " For there are three that bear record in
heaven, the Father, the Son, and the Holy Ghost; and
these three are one." This text he considers as a gross
corruption of Scripture, which had its origin among the
Latins, who interpreted the Spirit, Water, and Blood, to
be the Father, Son, and Holy Ghost, in order to prove
them one. With the same view Jerome inserted the

Trinity in express words in his version. The Latins marked his variations in the margins of their books; and in the twelfth and following centuries, when the disputations of the schoolmen were at their height, the variation began to creep into the text in transcribing. After the invention of printing, it crept out of the Latin into the printed Greek, contrary to the authority of all the Greek manuscripts and ancient versions; and from the Venetian press it went soon after into Greece. After proving these positions, Sir Isaac gives the following paraphrase of this remarkable passage, which is given in italics.

" *Who is he that overcometh the world, but he that believeth that Jesus is the Son of God*, that Son spoken of in the Psalms, where he saith, 'Thou art my Son; this day have I begotten thee.' *This is he that*, after the Jews had long expected him, *came*, first in a mortal body, *by* baptism of *water, and* then in an immortal one, by shedding his *blood* upon the cross, and rising again from the dead; *not by water only, but by water and blood*; being the Son of God, as well by his resurrection from the dead, (Acts xiii. 33,) as by his supernatural birth of the Virgin. (Luke i. 35.) *And it is the Spirit* also *that*, together with the water and blood, *beareth witness* of the truth of his coming; *because the Spirit is truth*; and so a fit and unexceptionable witness. *For there are three that bear record* of his coming; *the Spirit*, which he promised to send; and which was since shed forth upon us in the form of cloven tongues, and in various gifts; *the* baptism of *water*, wherein God testified, 'This is my beloved Son;' *and the* shedding of his *blood*, accompanied with his resurrection, whereby he became the most faithful martyr, or witness, of this truth. *And these three*, the spirit, the baptism, and passion of Christ, *agree in* witnessing *one* and the same thing; (namely, that the Son of God is come;) and, therefore, their evidence is strong: for the law requires but two consenting witnesses, and here we have three: *and if we receive the witness of men, the* threefold *witness of God,* which he bare of his Son, by declaring at his baptism 'This is my beloved Son,' by raising him from the dead, and by pouring out his Spirit on us, *is greater;* and, therefore, ought to be more readily received."

The other notable corruption of Scripture discussed by Sir Isaac is that which he charges the Greeks with having perpetrated in the text of St. Paul's First Epistle to Timothy, ch. iii. v. 16.: *Great is the Mystery of Godliness, God manifest in the Flesh.* According to him, this reading was effected by changing ΟΣ into ΘΣ, the abbreviation of Θεος; and, after proving this by a learned and ingenious examination of ancient manuscripts,* he concludes that the reading should be, *Great is the mystery of Godliness who* (viz., our Saviour) *was manifest in the flesh.*

As the tendency of Newton's Treatise was to deprive the defenders of the doctrine of the Trinity of the aid of two leading texts, he has been regarded by the Socinians and Arians, and even by some orthodox divines, as an Antitrinitarian; but such a conclusion is not warranted

* So long as there was any doubt whether the manuscript known as the Codex Alexandrinus read ΘΟ (the received contraction for ΘΕΟΣ, in cursive Greek Θεος), it was still somewhat uncertain what was the true reading of this celebrated text. Recently, however, this point has been set at rest. The late Dean Alford, in the last edition of his Greek Testament, informs us that he satisfied himself by an examination of that Codex at the British Museum that the supposed interior stroke converting Ο into Θ, was really that of an ε seen through from the other side of the leaf; and that Professor Maskelyne had ascertained, by putting the word under a powerful microscope, that there was neither a stroke within the Ο nor a mark of contraction over the word, so that the latter is undoubtedly ΟΟ, as in the Sinaitic and Cambridge manuscripts. (The Vatican MS. does not contain the Pastoral Epistles.) This therefore is the correct reading, in cursive Greek ὃς, i.e. the relative pronoun *Who*, referring to the antecedent understood, λόγος or Χριστος, Christ, the great "mystery of godliness," as the passage is rightly explained by Cyril of Alexandria, in his book *De Fide ad Imperatorem*, pertinently quoted by Newton.—EDITOR.

by any thing which he has published;[*] and he distinctly
warns us, that his object was solely to "purge the truth
of things spurious." We are disposed, on the contrary,
to think that he declares his belief in the doctrine of the
Trinity when he says, "In the eastern nations, and for a
long time in the western, the *faith* subsisted without this
text; and it is rather a danger to religion than an advantage, to make *it now* lean upon a bruised reed. There
cannot be better service done to the truth, than to purge
it of things spurious; and, therefore, knowing your prudence and calmness of temper, I am confident I shall not
offend you by telling you my mind plainly; especially
since it is no article of faith, no point of discipline,
nothing but a criticism concerning a text of scripture,
which I am going to write about." The word "faith"
in the preceding passage cannot mean faith in the scriptures in general, but faith in the particular doctrine of the
Trinity, for it is this article of faith only to which the
author refers when he deprecates *its* leaning on a bruised
reed. But, whatever be the meaning of this passage, we
know that Sir Isaac was greatly offended at Mr. Whiston
for having represented him as an Arian; and so much
did he resent the conduct of his friend in ascribing to him
heretical opinions, that he would not permit him to be
elected a Fellow of the Royal Society while he was
President.[†]

The only other religious works which were composed
by Sir Isaac Newton were his *Lexicon Propheticum*, to
which was added a Dissertation on the sacred cubit of the
Jews, and *Four Letters addressed to Dr. Bentley, containing some Arguments in proof of a Deity.*

[*] M. Biot has well remarked, that there is absolutely nothing in the
writings of Newton to justify, or even to authorize, the like, that he was
an Antitrinitarian. This passage is strangely omitted in the English
translation of Biot's Life of Newton.

[†] Whiston's Memoirs of his own Life, pp. 178, 249, 250. Edit. 1753.

The *Lexicon Propheticum* was left incomplete, and has not been published; but the Latin Dissertation which was appended to it, in which he shows that the cubit was about 26¼ Roman unciæ, was published in 1737 among the Miscellaneous Works of Mr. John Greaves.

Upon the death of the Hon. Robert Boyle on the 30th of December, 1691, it was found, by a codicil to his will, that he left a revenue of £50 per annum, to establish a lectureship, in which eight discourses were to be preached annually in one of the churches of the metropolis, in illustration of the evidences of Christianity, and in opposition to the principles of infidelity. Dr. Bentley, though a very young man, was appointed to preach the first course of sermons, and the manner in which he discharged this important duty gave the highest satisfaction not only to the trustees of the lectureship, but to the public in general. In the first six lectures Bentley exposed the folly of atheism even in reference to the present life; and derived powerful arguments for the existence of a Deity from the faculties of the soul, and the structure and functions of the human frame. In order to complete his plan, he proposed to devote his seventh and eighth lectures to the demonstration of a Divine Providence from the physical constitution of the universe, as established in the Principia. In order to qualify himself for this task, he received from Sir Isaac written directions respecting a list of books necessary to be perused previous to the study of that work; * and having made himself master of the system which it contained, he applied it with irresistible force of argument to establish the existence of an overruling mind. Previous to the publication of these lectures, Bentley encountered a difficulty which he was not able to solve, and he prudently transmitted to Sir Isaac, during 1692, a series of queries on the subject.

* Dr. Monk's *Life of Bentley*, p. 31.

This difficulty occurred in an argument urged by Lucretius, to prove the eternity of the world from a hypothesis of deriving the frame of it by mechanical principles from matter endowed with an innate power of gravity, and evenly scattered throughout the heavens. Sir Isaac willingly entered upon the consideration of the subject, and transmitted his sentiments to Dr. Bentley in the four letters which have been noticed in a preceding chapter.

In the first* of these letters Sir Isaac mentions that when he wrote his treatise about our system, viz., the Third Book of the Principia, " he had an eye upon such principles as might work, with considering men, for the belief of a deity," and he expresses his happiness that it has been found useful for that purpose. In answering the first query of Dr. Bentley, the exact import of which we do not know, he states, that, if matter were evenly diffused through a finite space, and endowed with innate gravity, it would fall down into the middle of the space, and form one great spherical mass ; but if it were diffused through an infinite space, some of it would collect into one mass, and some into another, so as to form an infinite number of great masses. In this manner the sun and stars might be formed if the matter were of a lucid nature. But he thinks it inexplicable by natural causes, and to be ascribed to the counsel and contrivance of a voluntary agent, that the matter should divide itself into two sorts, part of it composing a shining body like the sun, and part opaque bodies like the planets. Had a natural and blind cause, without contrivance and design, placed the earth in the centre of the moon's orbit, and Jupiter in the centre of his system of satellites, and the sun in the centre of the planetary system, the sun would have been a body like

* Dated December 10th, 1692. This letter is endorsed in Bentley's hand : "Mr. Newton's answer to some queries sent by me after I had preached my two last sermons."—Monk's *Life of Bentley*, p. 84, note.

Jupiter and the earth, that is, without light and heat; and consequently, he knows no reason why there is only one body qualified to give light and heat to all the rest, but because the author of the system thought it convenient, and because one was sufficient to warm and enlighten all the rest.

To the second query of Dr. Bentley he replies, that the motions which the planets now have could not spring from any natural cause alone, but were impressed by an intelligent agent. "To make such a system with all its motions, required a cause which understood, and compared together the quantities of matter in the several bodies of the sun and planets, and the gravitating powers resulting from thence; the several distances of the primary planets from the sun, and of the secondary ones from Saturn, Jupiter, and the earth, and the velocities with which those planets could revolve about those quantities of matter in the central bodies; and to compare and adjust all these things together in so great a variety of bodies, argues that cause to be not blind and fortuitous, but very well skilled in mechanics and geometry."

In the second* letter, he admits that the spherical mass formed by the aggregation of particles would affect the figure of the space in which the matter was diffused, provided the matter descends directly downwards to that body, and the body has no diurnal rotation; but he states, that by earthquakes loosening the parts of this solid, the protuberance might sink a little by their weight, and the mass by degrees approach a spherical figure. He then proceeds to correct an error of Dr. Bentley in supposing that all infinites are equal.—He admits that gravity might put the planets in motion; but he maintains that, without the Divine power, it could never give them such

* Dated Jan. 17th, 1692-3.

a circulating motion as they have about the sun, because a proper quantity of a transverse motion is necessary for this purpose; and he concludes that he is compelled to ascribe the frame of this system to an intelligent agent.

The third letter contains opinions confirming or correcting several positions which Dr. Bentley had laid down, and he concludes it with a curious examination of the opinion of Plato, that the motion of the planets is such as if they had been all created by God in some region very remote from our system, and let fall from thence towards the sun, their falling motion being turned aside into a transverse one whenever they arrived at their several orbits. Sir Isaac shows that there is no common place such as that conjectured by Plato, provided the gravitating power of the sun remains constant; but that Plato's affirmation is true, if we suppose the gravitating power of the sun to be doubled at that moment of time when they all arrive at their several orbits. "If we suppose," says he, "the gravity of all the planets towards the sun to be of such a quantity as it really is, and that the motions of the planets are turned upwards, every planet will ascend to twice its height from the sun. Saturn will ascend till he be twice as high from the sun as he is at present, and no higher; Jupiter will ascend as high again as at present, that is, a little above the orb of Saturn ; Mercury will ascend to twice his present height, that is, to the orb of Venus; and so of the rest; and then, by falling down again from the places to which they ascended, they will arise again at their several orbs with the same velocities they had at first, and with which they now revolve.

"But if so soon as their motions by which they revolve are turned upwards, the gravitating power of the sun, by which their ascent is perpetually retarded, be diminished by one-half, they will now ascend perpetually, and all of them, at all equal distances from the sun, will be equally swift. Mercury, when he arrives at the orb of Venus, will be as swift as Venus; and he and Venus, when they arrive at the orb of the earth, will be as swift as the earth; and so of the rest. If they begin all of them to ascend at once, and ascend in the same line, they will constantly, in ascending, become nearer and nearer together, and their motions will constantly approach to an equality, and become at length slower than any motion assignable. Suppose, therefore, that they ascended till they were almost contiguous, and their motions inconsiderably little, and that all their motions were at the same moment of time turned back again, or, which comes almost to the same thing, that they were only deprived of their motions, and let fall at that time, they would all at once arrive at their several orbs, each with the velocity it had at first; and if their motions were then turned sideways, and at the same time the gravitating power of the sun doubled, that it might be strong enough to retain them in their orbs, they would revolve in them as before their ascent. But if the gravitating power of the sun was not doubled, they would go away from their orbs into the highest heavens in parabolical lines."[*]

In the fourth letter[†] he states, that the hypothesis that matter is at first evenly diffused through the universe, is in his opinion inconsistent with the hypothesis of innate gravity without a supernatural power to reconcile them, and therefore it infers a Deity. "For if there be innate gravity, it is impossible now for the matter of the earth and all the planets and stars to fly up from them, and become evenly spread throughout all the heavens without a supernatural power; and certainly that which can never be hereafter without a supernatural power, could never be heretofore without the same power."

These letters, of which we have endeavoured to give a

[*] "These things," says he, "follow from my *Princip. Math.* lib. i, prop. 33. 34, 35, 36."

[†] Dated February 11th, 1692.

brief summary, will well repay the most attentive perusal by the philosopher as well as the divine. They are written with much perspicuity of language, and great power of thought, and they contain results which incontestably prove that their author was fully master of his noblest faculties, and comprehended the profoundest parts of his own writings.[*]

The logical acuteness, the varied erudition, and the absolute freedom from all prejudice which shine throughout the theological writings of Newton, might have protected them from the charge of having been written in his old age, and at a time when a failure of mind was supposed to have unfitted him for his mathematical investigations. But it is fortunate for his reputation, as well as for the interests of Christianity, that we have been able to prove the incorrectness of such insinuations, and to exhibit the most irrefragable evidence that *all the theological writings* of Newton were composed in the vigour of his life, and before the crisis of that bodily disorder which is supposed to have affected his reason. The able letters to Dr. Bentley were even written in the middle of that period when want of sleep and appetite had disturbed the serenity of his mind, and enable us to prove that this disturbance, whatever was its amount, never affected the higher functions of his understanding.

When a philosopher of distinguished eminence, and we believe not inimical to the Christian faith, has found it necessary to make a laboured apology for a man like

* The originals of these four letters to Bentley " were given by Dr. Richard Bentley to Cumberland, his nephew and executor, while a student at Trinity College, and were printed by him in a separate pamphlet in 1756. This publication was reviewed by Dr. Samuel Johnson in the Literary Magazine, vol. i, p. 89. See Johnson's Works, vol. ii, p. 528. The original letters are preserved in Trinity College, to which they were given by Cumberland a short time before his death."—Monk's *Life of Bentley,* p. 88, note.

Newton writing on theological subjects, and has been led
to render that apology more complete by referring this
class of his labours to a mind debilitated by age and
weakened by its previous aberrations, it may be expected
from an English biographer, and one who acknowledges
the importance of revealed truth, and the paramount interests of such subjects above all secular studies, to suggest
the true origin of Newton's theological inquiries.

When a mind of great and acknowledged power first
directs its energies to the study of the material universe,
no indications of order attract his notice, and no proofs
of design call forth his admiration. In the starry firma-
ment he sees no bodies of stupendous magnitude, and no
distances of immeasurable span. The two great luminaries
appear vastly inferior in magnitude to many objects around
him, and the greatest distances in the heavens seem even
inferior to those which his own eye can embrace on the
surface of the earth. The planets, when observed with
care, are seen to have a motion among the fixed stars, and
to vary in their magnitudes and distances, but these changes
appear to follow no law. Sometimes they move to the
east, sometimes to the west, passing the meridian sometimes near and sometimes far from the horizon, while at
other times they are absolutely stationary in their paths.
No system, in short, appears, and no general law seems
to direct their motions. By the observations and inquiries
of astronomers, however, during successive ages, a regular
system has been recognized in this chaos of moving bodies,
and the magnitudes, distances, and revolutions of all the
planets which compose it have been determined with the
most extraordinary accuracy. Minds fitted and prepared
for this species of inquiry are capable of appreciating the
great variety of evidence by which the facts of the planetary system are established; but thousands of individuals,
and many who are highly distinguished in other branches

of knowledge, are incapable of understanding such researches, and view with a sceptical eye the great and irrefragable truths of astronomy.

That the sun is stationary in the centre of our system,—that the earth moves round the sun, and round its own axis,—that the diameter of the earth is nearly 8000 miles, and that of the sun *one hundred and ten* times as great ; that the earth's orbit is more than 180 millions of miles in breadth ;—and that, if this immense space were filled with light, it would appear only like a luminous point at the nearest fixed star,—are positions absolutely unintelligible and incredible to all who have not carefully studied the subject. To millions of our species, then, the Great Book of Nature is absolutely sealed, though it is in the power of all to unfold its pages, and to peruse those glowing passages which proclaim the power and wisdom of its Author.

The Book of Revelation exhibits to us the same peculiarities as that of Nature. To the ordinary eye it presents no immediate indications of its divine origin. Events apparently insignificant—supernatural interferences seemingly unnecessary—doctrines almost contradictory—and prophecies nearly unintelligible—occupy its pages. The history of the fall of man—of the introduction of moral and physical evil—the prediction of a Messiah—the advent of our Saviour—his precepts—his miracles—his death—his resurrection—the gift of tongues—and the subsequent propagation of his religion by the unlettered fishermen of Galilee, are each a stumbling-block to the wisdom of this world. The youthful and vigorous mind, when summoned from its early studies to peruse the Scriptures, turns from them with disappointment. It recognizes in them no profound science—no secular wisdom—no divine eloquence—no disclosures of Nature's secrets—no palpable impress of an Almighty hand. But, though the system of revealed truth which the Scriptures contain is, like that of the universe, concealed from common observation, yet the labours of centuries have established its divine origin, and developed in all its order and beauty the great plan of human restoration. In the chaos of its incidents, we discover the whole history of our species, whether it is delineated in events that are past, or shadowed forth in those which are to come,—from the creation of man and the origin of evil, to the extinction of his earthly dynasty, and the commencement of his immortal career.

The antiquity and authenticity of the books which compose the sacred canon,—the fulfilment of its prophecies,—the miraculous works of its founder,—his death and resurrection, have been demonstrated to all who are capable of appreciating the force of historical evidence; and in the poetical and prose compositions of the inspired authors, we discover a system of doctrine, and a code of morality, traced in characters as distinct and legible as the most unerring truths in the material world. False systems of religion have indeed been deduced from the secret record, as false systems of the universe have sprung from the study of the book of nature; but the very prevalence of a false system proves the existence of one that is true; and though the two classes of facts necessarily depend on different kinds of evidence, yet we scruple not to say that the Copernican system is not more demonstrably true than the system of theological truth contained in the Bible. If men of high powers, then, are still found, who are insensible to the evidence which has established the system of the universe, need we wonder that there are others whose minds are shut against the effulgent evidence which sustains the strongholds of our faith?

If such is the character of the Christian Faith, we need not be surprised that it was embraced and expounded by such a genius as Sir Isaac Newton. Cherishing its doctrines, and leaning on its promises, he felt it his duty, as it was his delight, to apply to it that intellectual strength which had successfully surmounted the difficulties of the material universe. The fame which that success procured him he could not but feel to be the breath of popular applause, which administered only to his personal feelings ; but the investigation of the sacred mysteries, while it prepared his own mind for its final destiny, was calculated to promote the spiritual interests of thousands. This noble impulse he did not hesitate to obey, and by thus uniting philosophy with religion, he dissolved the league which genius had formed with scepticism, and added to the cloud of witnesses the brightest name of ancient or of modern times.

> " What wonder then that his devotion swelled
> Responsive to his knowledge! for could he,
> Where piercing mental eye diffusive saw
> The finished university of things,
> In all its order, magnitude, and parts,
> Forbear incessant to adore that Power
> Which fills, sustains, and actuates the whole?"—THOMSON.

CHAPTER XVII.

The Minor Discoveries and Inventions of Newton—His Researches on Heat—On Fire and Flame—On Elective Attraction—On the Structure of Bodies—His supposed Attachment to Alchemy—His Hypothesis respecting Æther as the Cause of Light and Gravity—On the Excitation of Electricity in Glass—His Reflecting Sextant invented before 1700—His Reflecting Microscope—His Prismatic Reflector as a Substitute for the Small Speculum of Reflecting Telescopes—His Method of Varying the Magnifying Power of Newtonian Telescopes—His Experiments on Impressions on the Retina.

IN the preceding chapters, we have given an account of the principal labours of Sir Isaac Newton; but there still remain to be noticed several of his minor discoveries and inventions, which could not properly be introduced under any general head.

The most important of these, perhaps, are his chemical researches, which he seems to have pursued with more or less diligence from the time when he first witnessed the practical operations of chemistry during his residence at the apothecary's at Grantham. His first chemical experiments were probably made on the alloys of metals, for the purpose of obtaining a good metallic composition for the specula of reflecting telescopes. In his paper on thin plates, he treats of the combinations of solids and fluids; but he enters more largely on these and other subjects in the queries published at the end of his Optics.

The only chemical paper of importance published by Sir Isaac, was read at the Royal Society on the 28th of May, 1701, and printed in the *Philosophical Transactions* without his name, under the title of *Scala graduum caloris*. This short paper contains a comparative scale of temperatures from that of melting ice to that of a small kitchen coal fire. The following are the principal points of the scale, the intermediate degrees of heat having been determined with great care.

Degrees of Heat.	Equal Parts of Heat.	
0	0	Freezing point of water.
1	12	Blood heat.
2	24	Heat of melting wax.
8	48	Melting point of equal parts of tin and bismuth.
4	96	Melting point of lead.
5	192	Heat of a small coal fire.

The first column of this table contains the degrees of heat in arithmetical progression; and the second contains the degrees of heat in geometrical progression, the second degree being twice as great as the first, and so on. It is obvious from this table, that the heat at which equal parts of tin and bismuth melt is *four* times greater than that of blood heat; the heat of melting lead *eight* times greater; and the heat of a small coal fire *sixteen* times greater.

This table was constructed by the help of a thermometer, and of red-hot iron. By the former he measured all heats as far as that of melting tin; and by the latter he measured all the higher heats. For the heat which heated iron loses in a given time is as the total heat of the iron; and, therefore, if the times of cooling are taken equal, the heats will be in a geometrical progression, and may therefore be easily found by a table of logarithms.

He found by a thermometer constructed with linseed oil, that if the oil, when the thermometer was placed in melting snow, occupied a space of 10,000 parts, the same oil, rarefied with *one* degree of heat, or that of the human body, occupied a space of 10,256; in the heat of water beginning to boil, a space of 10,705; in the heat of water boiling violently, 10,725; in the heat of melted tin beginning to cool, and putting on the consistency of an amalgam, 11,516, and when the tin had become solid, 11,496. Hence the oil was rarefied in the ratio of 40 to 39 by the heat of the human body; of 15 to 14 by the heat of boiling water; of 15 to 13 in the heat of melting tin beginning to solidify; and of 23 to 20 in the same tin when solid. The rarefaction of air was, with the same heat, *ten* times greater than that of oil; and the rarefaction of oil *fifteen* times greater than that of spirit of wine. By making the heat of oil proportional to its rarefaction, and by calling the heat of the human body 12 parts, we obtain the heat of water beginning to boil, 33; of water boiling violently, 34; of melted tin beginning to solidify, 72; and of the same become solid, 70.

Sir Isaac then heated a sufficiently thick piece of iro—till it was red hot; and having fixed it in a cold place, where the wind blew uniformly, he put upon it particles of different metals and other fusible bodies, and noted the times of cooling, till all the particles having lost their fluidity grew cold, and the heat of the iron was equal to that of the human body. Then, by assuming that the excesses of the heats of the iron and of the solidified particles of metal, above the heat of the atmosphere, were in geometrical progression when the times were in arithmetical progression, all the heats were obtained. The iron was placed in a current of air, in order that the air heated by the iron might always be carried away by the wind, and that cold air might replace it with a uniform motion; for thus equal parts of the air were heated in equal times, and received a heat proportional to that of the iron. But the heats thus found had the same ratio to one another with the heats found by the thermometer; and hence he was right in assuming, that the rarefactions of the oil were proportional to its heats.

The only other chemical paper of Newton that has been published is one of about two pages, entitled *De Natura Acidorum*, and is followed by other two pages, containing a number of brief opinions on chemical subjects. This paper was written later than 1687, as it bears a reference to the Principia; and the most important facts which it contains seem to have been more distinctly and fully reproduced in the queries at the end of the Optics.

The most interesting of these chemical queries relate to fire, flame, vapour, heat, and elective attractions; and as they were revised in 1716 and 1717, they may be regarded as containing the most mature opinions of their author. He regards fire as a body heated so hot as to emit light copiously; and flame as a vapour, fume, or exhalation, heated so hot as to shine.

In his long query on elective attractions, he considers the small particles of bodies as acting upon one another at distances so minute as to escape observation. When salt of tartar deliquesces, he supposes that this arises from an attraction between the saline particles and the aqueous particles held in solution in the atmosphere, and to the same attraction he ascribes it that the water will not distil from the salt of tartar without great heat. For the same reason sulphuric acid attracts water powerfully, and parts with it with great difficulty. When this attractive force becomes very powerful, as in the union between sulphuric acid and water, so as to make the particles "coalesce with violence," and rush towards one another with an accelerated motion, heat is produced by the mixture of the two

fluids. In like manner, he explains the production of
flame from the mixture of cold fluids,—the action of ful-
minating powders,—the combination of iron filings with
sulphur,—and all the other chemical phenomena of preci-
pitation, combination, solution, and crystallization, and
the mechanical phenomena of cohesion and capillary
attraction. He ascribes hot springs, volcanoes, fire-damps,
mineral coruscations, earthquakes, hot suffocating exhala-
tions, hurricanes, lightning, thunder, fiery meteors, sub-
terraneous explosions, land-slips, ebullitions of the sea,
and waterspouts, to sulphureous steams abounding in the
bowels of the earth, and fermenting with minerals, or
escaping into the atmosphere, where they ferment with
acid vapours fitted to promote fermentation.

In explaining the structure of solid bodies, he is of
opinion, " that the smallest particles of matter may cohere
by the strongest attractions, and compose bigger particles
of weaker virtue; and many of these may cohere and
compose bigger particles whose virtue is still weaker; and
so on for divers successions, until the progression end in
the biggest particles, on which the operations in chemistry,
and the colours of natural bodies, depend, and which, by
adhering, compose bodies of a sensible magnitude. If the
body is compact, and bends or yields inward to pressure,
without any sliding of its parts, it is hard and elastic,
returning to its figure with a force rising from the mutual
attraction of its parts. If the parts slide upon one another,
the body is malleable or soft. If they slip easily, and are
of a fit size to be agitated by heat, and the heat is big
enough to keep them in agitation, the body is fluid; and
if it be apt to stick to things, it is humid; and the drops
of every fluid affect a round figure, by the mutual attraction
of their parts, as the globe of the earth and sea affects a
round figure, by the mutual attraction of its parts by
gravity."

Sir Isaac then supposes, that, as the attractive force of bodies can reach but a small distance from them, "a repulsive virtue ought to succeed;" and he considers such a virtue as following from the reflexion and inflexions of the rays of light, the rays being repelled by bodies in both these cases without the immediate contact of the reflecting or inflecting body, and also from the emission of light, the ray, as soon as it is shaken off from a shining body by the vibrating motion of the parts of the body, getting beyond the reach of attraction, and being driven away with exceeding great velocity by the force of reflexion, the force that turns it back in reflexion being sufficient to emit it.*

Many of the chemical views which Newton thus published in the form of queries were in his own lifetime illustrated and confirmed by Dr. Stephen Hales, in his book on *Vegetable Statics*,—a work of great originality, which contains the germ of some of the finest discoveries in modern chemistry.

Although there is no reason to suppose that Sir Isaac Newton was a believer in the doctrines of alchemy, yet we are informed by the Rev. Mr. Law, that he had been a diligent student of Jacob Behmen's writings, and that there were found among his papers copious abstracts from them in his own handwriting.†   He states also that Sir

* Sir John Herschel, in his Treatise on Light, § 553, (Encyclopædia Metropolitana) has maintained that Newton's Doctrine of Reflexion is accordant with the idea, that the attractive force extends beyond the repulsive or reflecting force. In the query above referred to, Sir Isaac, in the most distinct manner, places the sphere of the reflecting force without that of the attractive one.

† In a tract annexed to his *Appeal to all that doubt or disbelieve the Truths of the Gospel.* See *Gent. Mag.*, 1782, vol. lii., pp. 227, 239.

It is stated in a letter of Mr. Law's quoted in this magazine that Charles I. was a diligent reader and admirer of Jacob Behmen; that he sent a well qualified person from England to Goerlitz, in Upper Lusatia, to acquire the German language, and to collect every anecdote he could meet with there relative to this great alchemist.

Isaac, together with one Dr. Newton, his relation, had, in the earlier part of his life, set up furnaces, and were for several months at work in quest of the philosopher's tincture. These statements may receive some confirmation from the fact, that there exist among the Portsmouth papers many sheets in Sir Isaac's own writing, of Flammel's Explication of Hieroglyphic Figures; and in another hand, many sheets of William Yworth's *Processus Mysterii magni Philosophicus*, and also from the manner in which Sir Isaac requests Mr. Ashton to inquire after one Borry in Holland, who always went clothed in green, and who was said to possess valuable secrets; but Mr. Law has weakened the force of his own testimony, when he asserts that Newton borrowed the doctrine of attraction from Behmen's first three propositions of eternal nature.

On the 7th of December, 1675, Sir Isaac Newton communicated to the Royal Society a paper entitled, *An Hypothesis explaining the Properties of Light*, in which he, for the first time introduces his opinions respecting ether, and employs them to explain the nature of light and the cause of gravity. " He was induced," he says, " to do this, because he had observed the heads of some great virtuosos to run much upon hypotheses, and he therefore gave one which he was inclined to consider as the most probable, if he were obliged to adopt one." *

This hypothesis seems to have been afterwards a subject of discussion between him and Mr. Boyle, to whom he promised to communicate his opinion more fully in writing. He accordingly addressed to him a long letter, dated February 28th, 1678–9, in which he explains his views respecting ether, and employs them to account for the refraction of light,—the cohesion of two polished pieces

* In a letter to Dr. Halley, dated June 20th, 1686, Sir Isaac refers to this paper, and observes, that it is only to be looked upon as one of his guesses that he did not rely upon.

of metal in an exhausted receiver,—the adhesion of quicksilver to glass tubes,—the cohesion of the parts of all
bodies,—the cause of filtration,—the phenomena of capillary attraction,—the action of menstrua on bodies,—the
transmutation of gross compact substances into aërial
ones,—and the cause of gravity. From the language used
in this paper, we should be led to suppose that Sir Isaac
had entirely forgotten that he had formerly treated the
general subject of ether, and applied it to the explanation
of gravity. "I shall set down," says he, "one conjecture
more *which came into my mind now as I was writing this
letter; it is about the cause of gravity,*" which he goes on
to explain ; and he concludes by saying, that he has "so
little fancy to things of this nature, that *had not your
encouragement moved me to it,* I should never, I think,
thus far have set pen to paper about them."

These opinions, however, about the existence of ether,
Newton seems to have subsequently renounced; for in the
manuscript in the possession of Dr. J. C. Gregory, which
we have already mentioned, and which was written previous to 1702, he states, that ether is neither obvious to our
senses, nor supported by any arguments, but is a gratuitous
assumption, which, if we are to trust to reason and to our
senses, must be banished from the nature of things ; and
he goes on to establish, by various arguments, the validity
of this opinion. This renunciation of his former hypothesis
probably arose from his having examined more carefully
some of the phenomena which he endeavoured to explain
by it. Those of capillary attraction, for example, he had
ascribed to the ether "standing rarer in the very sensible
cavities of the capillary tubes than without them," whereas
he afterwards discovered their true cause, and ascribed them
to the reciprocal attraction of the tube and the fluid. But,
however this may be, there can be no doubt that he resumed his early opinions before the publication of his

T

Optics, which may be considered as containing his views upon this subject.

The queries which contain these opinions are the 18th–24th, all of which appeared for the first time in the second English edition of the Optics. If a body is either heated or loses its heat when placed in vacuo, he ascribes the conveyance of the heat in both cases "to the vibration of a much subtiler medium than air;" and he considers this medium as the same with that by which light is refracted and reflected, and by whose vibrations light communicates heat to bodies, and is put into fits of easy reflexion and transmission.

This ethereal medium, according to our author, is exceedingly more rare and more elastic than air. It pervades all bodies, and is expanded through all the heavens. It is much rarer within the dense bodies of the sun, stars, planets, and comets, than in the celestial spaces between them, and also more rare within glass, water, &c., than in the free and open spaces void of air and other grosser bodies. In passing out of glass, water, &c., and other dense bodies, into empty space, it grows denser and denser by degrees, and this gradual condensation extends to some distance from the bodies. Owing to its great elasticity, and, consequently, its efforts to spread in all directions, it presses against itself, and, consequently, against the solid particles of bodies, so as to make them continually approach to one another, the body being impelled from the denser parts of the medium towards the rarer with all that power which we call gravity.

In employing this medium to explain the nature of light, Newton does not suppose with Descartes, Hooke, Huygens, and others, that light is nothing more than the impression of those undulations on the retina. He regards light as a peculiar substance composed of heterogeneous particles thrown off with great velocity, and in all directions, from luminous bodies; and he supposes that these particles, while passing through the ether, excite in it vibrations or pulses which accelerate or retard the particles of light, and thus throw them into their alternate fits of easy reflexion and transmission.

Hence, if a ray of light falls upon a transparent body, in which the ether consists of strata of variable density, the particles of light acted upon by the vibrations which they create will be urged with an accelerated velocity in entering the body, while their velocity will be retarded in quitting it. In this manner he conceives the phenomena of refraction to be produced, and he shows how in such a case the refraction would be regulated by the law of the sines.

In order that the ethereal medium may produce the fits of easy reflexion and transmission, he conceives that its vibrations must be swifter than light. He computes its elasticity to be 490,000,000,000 times greater than that of air in proportion to its density, and on the hypothesis that the latter is 600,000,000 times less than that of water, infers that the resistance which it would oppose to the motions of the planets would not be sensible in 10,000 years. He considers that the functions of vision and hearing may be performed chiefly by the vibrations of this medium, excited in the bottom of the eye by the rays of light, or in the auditory nerves by the tremors of the air, and propagated through the solid, pellucid, and uniform *capillamenta* of the optic or auditory nerves into the place of sensation; and thinks that animal motion may be performed by vibrations of the same medium, excited in the brain by the power of the will, and propagated from thence by the solid, pellucid and uniform *capillamenta* of the nerves into the muscles for contracting and dilating them.

In the registers of the Royal Society there exist several

letters* on the excitation of electricity in glass, which were occasioned by an experiment of this kind having been mentioned in Sir Isaac's hypothesis of light. The Society had ordered the experiment to be tried at their meeting of the 16th December, 1675; but, in order to secure its success, Mr. Oldenburg wrote to Sir Isaac for a more particular account of it. Sir Isaac, being thus "put upon recollecting himself a little farther about it," remembers that he made the experiment with a glass fixed at the distance of the $\frac{3}{8}$d of an inch from one end of a brass hoop, and only $\frac{1}{8}$th of an inch from the other. Small pieces of thin paper were then laid upon the table; when the glass was laid above them and rubbed, the pieces of paper leapt from the one part of the glass to the other, and twirled about in the air. Notwithstanding this explicit account of the experiment, it entirely failed at the Royal Society, and the secretary was desired to request the loan of Sir Isaac's apparatus, and to inquire whether or not he had secured the papers from being moved by the air, which might have somewhere stole in. In a letter, dated 21st of December, Sir Isaac recommended to the Society to rub the glass " with stuff whose threads may rake its surface, and, if that will not do, to rub it with the fingers' ends to and fro, and knock them as often upon the glass." These directions enabled the Society to succeed with the experiment on the 13th of January, 1676, when they used a scrubbing-brush of short hogs' bristles, and the heft of a knife made with whalebone.

Among the minor inventions of Sir Isaac Newton, we have already mentioned his reflecting instrument for observing the moon's distance from the fixed stars at sea. The description of this instrument was communicated to Dr. Halley in the year 1700; but, either from having

* See *Newtoni Opera*, by Horsley, vol. iv., pp. 375-382.

mislaid the manuscript, or from attaching no value to the invention, he never communicated it to the Royal Society, and it remained among his papers till after his death in 1742, when it was published in the Philosophical Transactions. The following is Sir Isaac's own description of it as communicated to Dr. Halley.

"In the annexed figure P Q R S denotes a plate of brass, accurately divided in the limb D Q, into $\frac{1}{2}$ degrees, $\frac{1}{2}$ minutes, and $\frac{1}{12}$ minutes, by a diagonal scale; and the $\frac{1}{2}$ degrees, and $\frac{1}{2}$ minutes, and $\frac{1}{12}$ minutes, counted for degrees, minutes, and $\frac{1}{6}$ minutes. A B is a telescope three or four feet long, fixed on the edge of that brass plate. G is a speculum fixed on the brass plate perpendicularly as near as may be to the object-glass of the telescope, so as to be

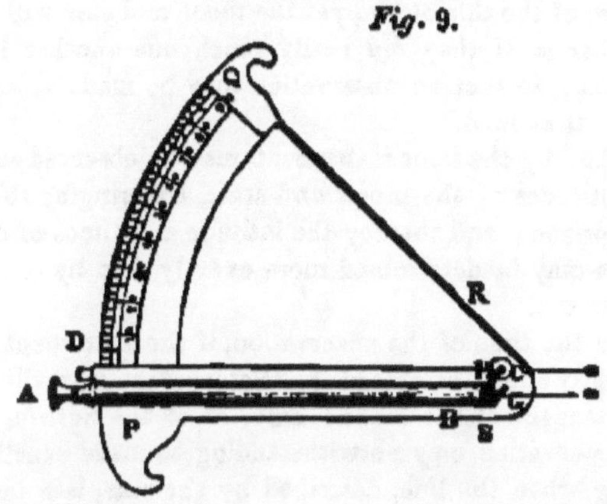

Fig. 9.

inclined forty-five degrees to the axis of the telescope, and intercept half the light which would otherwise come through the telescope to the eye. C D is a moveable index turning about the centre C, and, with its fiducial edge, showing the degrees, minutes, and $\frac{1}{6}$ minutes on the limb of the brass plate P Q; the centre C must be over against the middle of the speculum G. H is another speculum, parallel to the former, when the fiducial edge of index falls on 0° 0′ 0″; so that the same star may then appear through the telescope in one and the same place, both by the direct rays and by the reflexed ones; but if the index be turned, the star shall appear in two places, whose distance is showed on the brass limb by the index.

"By this instrument the distance of the moon from any fixed star is thus observed: view the star through the perspicil by the direct light, and the moon by the reflexed, (or on the contrary,) and turn the index till the star touch the limb of the moon, and the index shall show on the brass limb of the instrument the distance of the star from the moon's limb; and though the instrument shake by the motion of the ship at sea, yet the moon and star will move together as if they did really touch one another in the heavens; so that an observation may be made as exactly at sea as at land.

"And by the same instrument, may be observed exactly the altitudes of the moon and stars, by bringing them to the horizon; and thereby the latitude and times of observation may be determined more exactly than by the ways now in use.

"In the time of the observation, if the instrument move angularly about the axis of the telescope, the star will move in a tangent of the moon's limb, or of the horizon; but the observation may notwithstanding be made exactly, by noting when the line, described by the star, is a tangent to the moon's limb, or to the horizon.

"To make the instrument useful, the telescope ought to take in a large angle; and, to make the observation true, let the star touch the moon's limb, not on the outside, but on the inside."

This ingenious contrivance is obviously the very same

invention as that which Mr. Hadley produced in 1731, and which, under the name of Hadley's Sextant, has been of so great service in navigation. The merit of its first invention must therefore be transferred to Sir Isaac Newton.

In the year 1672, Sir Isaac communicated to Mr. Oldenburg his design for a microscope, which he considered to be as capable of improvement as the telescope, and perhaps more so, because it requires only one speculum. This microscope is shown in the annexed diagram, where

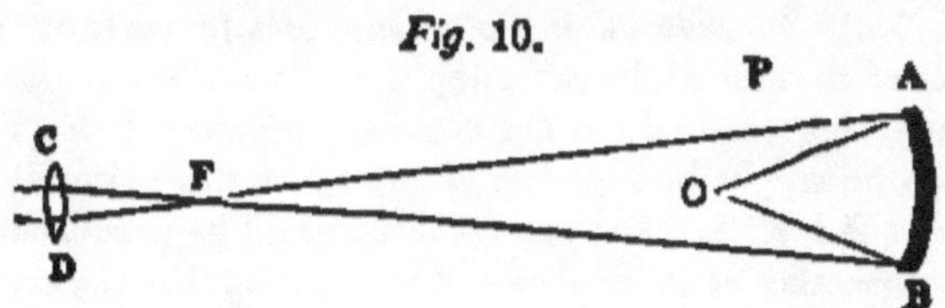

A B is the object-metal, C D the eye-glass, F their common focus, and O the other focus of the metal in which the object is placed. This ingenious idea has been greatly improved in modern times by Professor Amici, Professor Potter and Dr. Goring, who make A B a portion of an ellipsoid, whose foci are O and F, and who place a small plain speculum between O and A B, in order to reflect into the speculum the object, which is placed on one side A P, for the purpose of being illuminated.

In another letter to Mr. Oldenburg, dated July 11th in the same year, he suggests an improvement of microscopes by refraction which is to "illuminate the object in a darkened room with the light of any convenient colour not too much compounded: for by that means the microscope will, with distinctness, bear a deeper charge and larger aperture, especially if its construction be such as I may hereafter describe."[*] This happy idea we have some years ago succeeded in realizing, by illuminating microscopic objects with the light of a monochromatic lamp, which discharges a copious flame of pure yellow light of definite refrangibility.[†]

In order to remedy the evil of want of light in his reflecting telescope, arising from the weak reflecting power of speculum metal, and from its tarnishing by exposure to the air, Sir Isaac proposed to substitute for the small oval speculum a triangular prism of glass or crystal A B C (fig. 11.) Its side A B *b a* he supposes to perform the office of that metal, by reflecting towards the eye-glass the light which comes from the concave speculum D F (Fig. 12), whose light he supposes to enter into this prism at its side C B *b c;* and lest any colours should be produced by the refraction of these planes, it is requisite that the angles of the prism at A *a* and B *b* be precisely equal. This may be done most conveniently, by making them half right angles, and consequently the third angle at C *c* a right one. The plane A B *b a* will reflect all the light incident upon it; but in order to exclude unnecessary light, it will be proper to cover it all over with some black substance, excepting two circular spaces of the planes A *c* and B *c,* through which the useful light may pass. The length of the prism should be such that its sides A *c* and B *c* may be "four square," and so much of the angles B and *b* as are superfluous

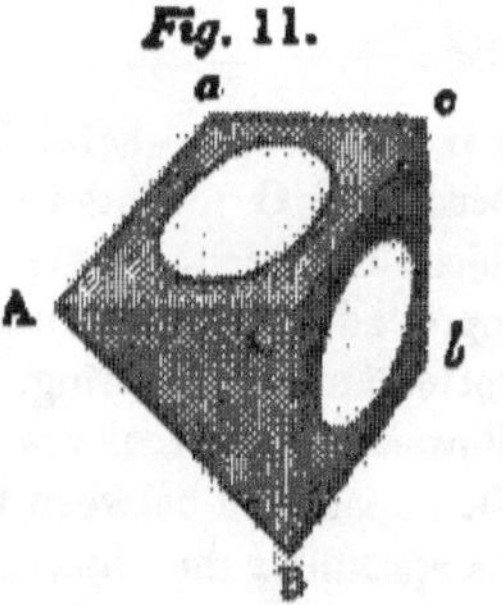

* Sir Isaac does not seem to have afterwards described this construction.

† See *Edinburgh Transactions*, vol. ix., p. 433, and the *Edinburgh Journal of Science*, July, 1829, No. I., New Series, p. 108.

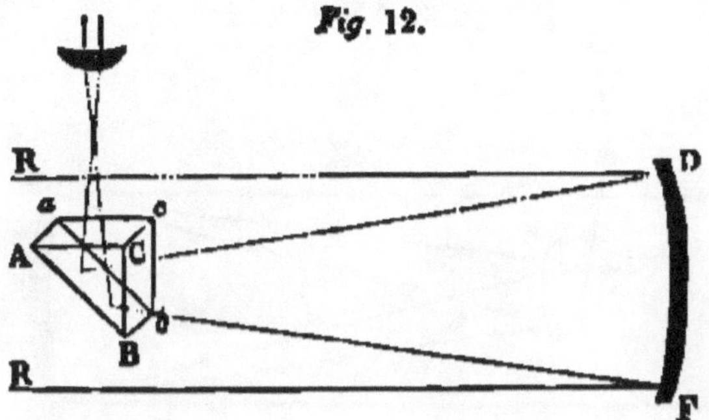

Fig. 12.

ought to be ground off, to give passage for as much light as is possible from the object to the speculum.

One great advantage of this prism, which cannot be obtained from the oval metal, is that, without using two glasses, the object may be erected, and the magnifying power of the telescope varied at pleasure, by merely varying the distances of the speculum, the prism, and the eyeglass. This will be understood from Fig. 13 (next page), where A I represents the great concave speculum, E F the eyeglass, and B C D the prism of glass, whose sides B C and B D are not flat, but spherically convex. The rays which come from G, the focus of the great speculum A I, will, by the refraction of the first side B D, be reduced to parallelism, and after reflexion from the base C D, will be made by the refraction of the next side B C to converge to the focus H of the eye-glass E F. If we now bring the prism B C D nearer the image at G, the point H will recede from B C, and the image formed there will be greater than that at G; and if we remove the prism B C D from G, the point H will approach to B C, and the image at H will be less than that at G. The prism B C D performs the same part as a convex lens, G and H being its conju-

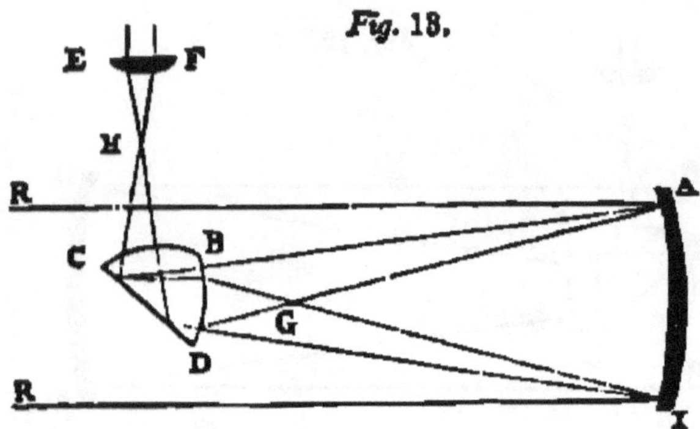

gate foci, and the relative size of the images formed at these points being proportional to their distance from the lens. This construction would be a good one for varying optically the angular distance of a pair of wires placed in the focus of the eye-glass E F; and by bisecting the lenticular prism B C D, and giving the halves a slight inclination, we should be able to separate and to close the two images or discs which would thus be produced, and thus form a double image micrometer.

Among the minor and detached labours of Sir Isaac, we must not omit his curious experiments on the action of light upon the retina. Locke seems to have wished his opinion respecting a fact stated in Boyle's Book on Colours; and in a letter from Cambridge, dated June 30th, 1691, he communicated to his friend the following very remarkable observations made by himself :—

" The observation you mention in Mr. Boyle's Book of Colours, I once made upon myself with the hazard of my eyes. The manner was this:—I looked a very little while upon the sun in the looking-glass with my right eye, and then turned my eyes into a dark corner of my chamber, and winked, to observe the impression made, and the circles

of colours which encompassed it, and how they decayed
by degrees, and at last vanished. This I repeated a second
and a third time. At the third time, when the phantasm
of light and colours about it were almost vanished, intend-
ing my fancy upon them to see their last appearance, I
found, to my amazement, that they began to return, and
by little and little to become as lively and vivid as when
I had newly looked upon the sun. But when I ceased to
intend my fancy upon them, they vanished again. After
this, I found, that, as often as I went into the dark, and
intended my mind upon them, as when a man looks
earnestly to see any thing which is difficult to be seen, I
could make the phantasm return without looking any more
upon the sun; and the oftener I made it return, the more
easily I could make it return again. And at length, by
repeating this without looking any more upon the sun, I
made such an impression on my eye, that if I looked upon
the clouds, or a book, or any bright object, I saw upon it
a round bright spot of light like the sun, and, which is
still stranger, though I looked upon the sun with my right
eye only, and not with my left, yet my fancy began to
make an impression upon my left eye, as well as upon my
right. For if I shut my right eye, or looked upon a book
or the clouds with my left eye, I could see the spectrum
of the sun almost as plain as with my right eye, if I did
but intend my fancy a little while upon it; for at first, if
I shut my right eye, and looked with my left, the spectrum
of the sun did not appear till I intended my fancy upon it;
but, by repeating, this appeared every time more easily.
And now, in a few hours' time, I had brought my eyes to
such a pass, that I could look upon no bright object with
either eye, but I saw the sun before me, so that I durst
neither write nor read; but, to recover the use of my eyes,
shut myself up in my chamber made dark, for three days
together, and used all means to divert my imagination

from the sun.   For if I thought upon him, I presently saw
his picture, though I was in the dark.   But by keeping in
the dark, and employing my mind about other things, I
began in three or four days to have some use of my eyes
again; and, by forbearing to look upon bright objects,
recovered them pretty well, though not so well but that,
for some months after, the spectrum of the sun began to
return as often as I began to meditate upon the phenomena,
even though I lay in bed at midnight with my curtains
drawn.   But now I have been very well for many years,
though I am apt to think, if I durst venture my eyes, I
could still make the phantasm return by the power of my
fancy.   This story I tell you, to let you understand, that
in the observation related by Mr. Boyle, the man's fancy
probably concurred with the impression made by the sun's
light, to produce the phantasm of the sun which he constantly saw in bright objects.   And so your question about
the cause of this phantasm involves another about the
power of fancy, which, I must confess, is too hard a knot
for me to untie.   To place this effect in a constant motion
is hard, because the sun ought then to appear perpetually.
It seems rather to consist in a disposition of the sensorium
to move the imagination strongly, and to be easily moved,
both by the imagination and by the light, as often as
bright objects are looked upon."

These observations possess in many respects a high
degree of interest.   The fact of the transmission of the
impression from the retina of the one eye to that of the
other, is particularly important; and it deserves to be remarked as a singular coincidence, that I had occasion to
observe and to describe the same phenomena above twenty
years ago,* and long before the observations of Sir Isaac
were communicated to the scientific world.

* Art. ACCIDENTAL COLOURS, in the *Edinburgh Encyclopædia.*

CHAPTER XVIII.

Newton's Acquaintance with Dr. Pemberton—Who edits the Third Edition of the Principia—His first Attack of ill health—His Recovery—He is taken Ill in Consequence of Attending the Royal Society—His Death on the 20th of March, 1727—His Body lies in State—His Funeral—He is Buried in Westminster Abbey—His Monument described—His Epitaph—A Medal struck in Honour of him—Roubiliac's full-length Statue of him erected in Cambridge—Division of his Property—His Successors.

ABOUT the year 1722, Sir Isaac was desirous of publishing a third edition of his Principia, for which ho had long been making preparations. The premature death of Mr. Cotes had deprived him of his valuable assistance, but he had the good fortune to obtain the services of Dr. Henry Pemberton, a young and accomplished physician, who had cultivated mathematical learning with considerable success. Signior Poleni, an Italian mathematician, having endeavoured, on the authority of a new experiment, to prove Leibnitz's assertion that the force of descending bodies is proportional to the square of the velocity, and not to the simple velocity, Dr. Pemberton transmitted to Dr. Mead a demonstration of its insufficiency. Dr. Mead communicated this paper to Sir Isaac, who was so well pleased with it, that he called upon Pemberton at his lodgings, and showed him a refutation of Poleni by himself, grounded on other principles. This was afterwards printed without his name, as a postscript to Pemberton's paper, which was published in the form of a letter to Dr. Mead, in the Philosophical Transactions for 1722. In this agreeable way, Pemberton secured what he had ardently longed for, the friendship of Newton, and they often met together to converse on mathematical and philosophical subjects.

In a short time after the commencement of their acquaintance, Sir Isaac engaged Dr. Pemberton to superintend the new edition of the Principia. Iu discharging this duty, Dr. Pemberton had occasion to make many suggestions for the improvement of the work, which, with one or two exceptions, seem to have been implicitly adopted by Newton.

In February or March, 1726, the third edition of the Principia was published, with a new preface by the author, in which he mentions the more important additions he had made, and states that Dr. Henry Pemberton,* "vir harum rerum peritissimus," superintended the edition. On the occasions upon which the latter had personal intercourse with Sir Isaac, and which were necessarily numerous, he endeavoured to learn his opinions on various mathematical subjects, and to obtain some historical information respecting his inventions and discoveries. Sir Isaac entered freely into all these topics; and during the conversations which took place, and while they were reading together Dr. Pemberton's popular account of Sir Isaac's discoveries, he obtained the most perfect evidence, that, though his memory was much decayed, yet he was fully able to understand his own writings.

During the last twenty years of his life, which he spent in London, the charge of his domestic concerns devolved upon his beautiful and accomplished niece, Mrs. Catharine Barton, the widow of Colonel Barton, for whom, as we

* Pemberton died in the year 1771, at the age of 77.—EDITOR.

have already seen, the Earl of Halifax had conceived the warmest affection. This lady, who had been educated at her uncle's expense, married Mr. Conduitt, and continued to reside with her husband in Sir Isaac's house till the time of his death.

In the year 1722, when he had reached the eightieth year of his age, he was seized with an incontinence of urine, which was ascribed to stone in the bladder, and was considered incurable. By means of a strict regimen, however, and other precautions, he was enabled to alleviate his complaint, and to procure long intervals of ease. At this time he gave up the use of his carriage, and always went out in a chair. He declined all invitations to dinner, and at his own house he had only small parties. In his diet he was extremely temperate. Though he took a little butcher's meat, yet the principal articles of his food were broth, vegetables, and fruit, of which he always ate very heartily. In spite of all his precautions, however, he experienced a return of his old complaint; and in August, 1724, he passed, without any pain, a stone about the size of a pea, which came away in two pieces, the one at an interval of some days from the other. After some months of tolerably good health, he was seized, in January, 1725, with a violent cough and inflammation of the lungs; and, in consequence of this attack, he was prevailed upon, with some difficulty, to take up his residence at Kensington, where his health experienced a decided improvement. In February, 1725, he was attacked in both his feet with a fit of the gout, of which he had received a slight warning a few years before, and the effect of this new complaint was to produce a beneficial change in his general health. On Sunday, the 7th of March, when his head was clearer and his memory stronger than Mr. Conduitt had known it to be for some time, he entered into a long conversation on various topics in astronomy of a speculative nature, which

Mr. Conduitt, who knew little of the subject, has, we think, very imperfectly reported.[*]

Notwithstanding the improvement which his health had experienced, his indisposition was still sufficiently severe to unfit him for the discharge of his duties at the Mint; and, as his old deputy was confined with the dropsy, he was desirous in 1725 of resigning in favour of Mr. Conduitt. Difficulties, however, seem to have been experienced in making this arrangement, but all the duties of the office were so satisfactorily performed by Mr. Conduitt that during the last year of his life, Sir Isaac hardly ever went to the Mint.

But, though every kind of motion was calculated to aggravate his complaint, and though he had derived from absolute rest, and from the air at Kensington, the highest benefit, yet great difficulty was experienced in preventing him from occasionally going to town. Thinking that he was fit for the journey, he went to London on Tuesday the 28th of February, 1727, to preside at a meeting of the Royal Society, on the 2nd of March, and on the following day Mr. Conduitt considered him better than he had been for many years. Sir Isaac himself was sensible of the change, and told his nephew, smiling, that on the Sunday preceding he had slept from eleven o'clock at night till eight in the morning, without waking. These feelings, however, were fallacious. He had undergone great fatigue in attending the meeting of the Society, and in paying and receiving visits, and the consequence of this was a violent attack of his former complaint. He returned to Kensington on Saturday the 4th of March, and was attended by Dr. Mead and Mr. Cheselden, who pronounced his disease to be stone in the bladder, and held out no hopes of his recovery. From the time of his last journey to London

---

[*] This conversation, originally copied from Mr. Conduitt's hand-writing is given in the Appendix, No. lii., p. 356.

he had experienced violent fits of pain with very brief intermissions; and though the drops of sweat ran down his face in these severe paroxysms, he never uttered a cry or a complaint, or displayed the least marks of peevishness or impatience; but during the short intervals of relief which occurred, " would smile and talk with his usual cheerfulness." On Wednesday the 15th of March he appeared to be somewhat better; and slight, though groundless, hopes were entertained of his recovery. On the morning of Saturday the 18th he read the newspapers, and carried on a pretty long conversation with Dr. Mead. His senses and faculties were still vigorous, but at six o'clock of the same evening, he became insensible, and continued in that state during the whole of Sunday and until Monday the 20th, when he expired without pain between one and two o'clock in the morning, in the eighty-fifth year of his age.

> .   .   .   'Tis done, the measure's full,
> And I resign my charge.—THOMSON.

His body was removed from Kensington to London, and on Tuesday, the 28th of March, it lay in state in the Jerusalem Chamber, and was thence conveyed to Westminster Abbey, where it was buried near the entrance into the choir on the left hand. The pall was supported by the Lord High Chancellor, the Dukes of Montrose and Roxburghe, and the Earls of Pembroke, Sussex, and Macclesfield, who were Fellows of the Royal Society. The Hon. Sir Michael Newton, Knight of the Bath, was chief mourner, and was followed by some other relations, and several eminent persons who were intimately acquainted with the deceased. The funeral service was performed by the Bishop of Rochester, attended by the prebends and choir.

Sensible of the high honour which they derived from their connexion with so distinguished a philosopher, the relations of Sir Isaac Newton who inherited his personal estate,* agreed to devote £500 to the erection of a monument to his memory, and the Dean and Chapter of Westminster appropriated for it a place in the most conspicuous part of the Abbey, which had often been refused to the greatest of our nobility. This monument was erected in 1731. On the front of a sarcophagus resting on a pedestal, are sculptured in basso relievo youths bearing in their hands the emblems of Sir Isaac's principal discoveries. One carries a prism, another a reflecting telescope, a third is weighing the sun and planets with a steelyard, a fourth is employed about a furnace, and two others are loaded with money newly coined. On the sarcophagus is placed the figure of Sir Isaac in a recumbent posture, with his elbow resting on several of his works. Two youths stand before him with a scroll, on which is drawn a remarkable diagram relative to the solar system, and above that is a converging series. Behind the sarcophagus is a pyramid, from the middle of which rises a globe in mezzo relievo, upon which several of the constellations are drawn, in order to show the path of the comet of 1681, whose period Sir Isaac had determined, and also the position of the solstitial colure mentioned by Hipparchus, and by means of which Sir Isaac had, in his Chronology, fixed the time of the Argonautic expedition. A figure of Astronomy as Queen of the Sciences sits weeping on the Globe with a sceptre in her hand, and a star surmounts the summit of the pyramid. The following epitaph is inscribed on the monument :—

* These were the three children of his half-brother Smith, the three children of his half-sister Pilkington, and the two daughters of his half-sister Barton, all of whom survived Sir Isaac.—*New Anecdotes of Sir Isaac Newton, by J. H., a Gentleman of his Mother's Family. See Annual Register*, 1776, vol., xix., p. 25, of Characters. The author of this paper was James Hutton, Esq., of Pimlico.

Hic situs est
Isaacus Newton, Eques Auratus,
Qui animi vi propè divinâ,
Planetarum Motus, Figuras,
Cometarum Semitas, Oceanique Æstus,
Suâ Mathesi facem præferente,
Primus demonstravit,
Radiorum Lucis dissimilitudines,
Colorumque inde nascentium proprietates,
Quas nemo antea vel suspicatus erat, pervestigavit.
Naturæ, Antiquitatis, S. Scripturæ,
Sedulus, sagax, fidus Interpres,
Dei Opt. Max. Majestatem philosophiâ asseruit,
Evangelii simplicitatem moribus expressit.
Sibi gratulentur Mortales, tale tantumque extitisse
HUMANI GENERIS DECUS.
Natus xxv. Decemb. MDCXLII, Obiit xx. Mart.
MDCCXXVII.

Of which the following is a literal translation :—

Here Lies
Sir Isaac Newton, Knight,
Who, by a vigour of mind almost supernatural,
First demonstrated, with the Torch of his Mathematics,
The Motions and Figures of the Planets,
The Paths of the Comets, and the Tides of the Ocean.
He diligently investigated
The different refrangibilities of the Rays of Light,
And the properties of the Colours to which they give rise,
Which no one had even imagined before.
An Assiduous, Sagacious, and Faithful Interpreter
of Nature, Antiquity, and the Holy Scriptures,
He asserted in his Philosophy the Majesty of God,
And exhibited in his Conduct the simplicity of the Gospel.
Let Mortals rejoice
That there has existed such and so great
AN ORNAMENT OF THE HUMAN RACE.
Born 25th Dec., 1642, Died 20th March, 1727.

In the beginning of 1731, a medal was struck at the
Tower in honour of Sir Isaac Newton. It had on one
side the head of the philosopher, with the motto, *Felix
cognoscere causas*, and on the reverse a figure representing
the mathematics.

On the 4th of July, 1755, a magnificent full-length statue of Sir Isaac Newton in white marble was erected in the ante-chapel of Trinity College.   He is represented standing on a pedestal in a loose gown, holding a prism, and looking upwards with an expression of the deepest thought.   On the pedestal is the inscription,—

> Qui genus humanum ingenio superavit.
> Who surpassed all men in genius.

This statue, executed by Roubilliac, was erected at the expense of Dr. Robert Smith, the author of the *Compleat System of Optics*, and Professor of Astronomy and Experimental Philosophy at Cambridge.—It has been thus described by a modern poet :—

> Hark where the organ, full and clear,
> With loud hosannahs charms the ear ;
> Behold, a prism within his hands,
> Absorbed in thought great Newton stands ;
> Such was his brow, and look serene,
> His serious gait and musing mien,
> When, taught on eagle wings to fly,
> He traced the wonders of the sky ;
> The chambers of the sun explored,
> Where tints of thousand hues were stored.

Dr. Smith likewise bequeathed the sum of £500 for executing a painting on glass for the window at the south end of Trinity College, Cambridge.   The subject represents the presentation of Sir Isaac Newton to his Majesty George III., who is seated under a canopy with a laurel chaplet in his hand, and attended by the British Minerva, apparently advising him to reward merit in the person of the great philosopher.   Below the throne, the Lord Chancellor Bacon is, by an anachronism legitimate in art, proposing to register the reward about to be conferred upon Sir Isaac.   The original drawing of this picture was executed by Cypriani, and cost one hundred guineas.

The personal estate of Sir Isaac Newton, which was worth about £32,000, was divided among his four nephews and four nieces of the half-blood, the grandchildren of his mother by the Rev. Mr. Smith. The family estates of Woolsthorpe and Sustern he bequeathed to John Newton, the heir-at-law, whose great-grandfather was Sir Isaac's uncle. This gentleman sold them in 1732 to Edmund Turnor of Stoke Rocheford. A short time before his death, Sir Isaac gave away an estate in the parish of Baydon, in Wiltshire, to the sons and daughter of a brother of Mrs. Conduitt, who, in consequence of their father dying before Sir Isaac, had no share in the personal estate; and he also gave an estate of the same value, which he bought at Kensington, to Catherine, the only daughter of Mr. Conduitt, who afterwards married Mr. Wallop. This lady was afterwards Viscountess Lymington, and the estate of Kensington descended to the Earl of Portsmouth, by whom it was sold. Sir Isaac was succeeded as Master in the Mint by his nephew, John Conduitt, Esq., who wrote a treatise on the gold and silver coin, and died in 1737, in the 49th year of his age. His widow, Mrs. Conduitt, erected a handsome monument to his memory in Westminster Abbey, and died in 1739, in the 59th year of her age.

CHAPTER XIX.

SUCH were the last days of Isaac Newton, and such the last laurels which were shed over his grave. A century of discoveries has, since his day, been added to science; but, brilliant as these discoveries are, they have not obliterated the minutest of his labours, and have served only to brighten the halo which encircles his name. The achievements of genius, like the source from which they spring, are indestructible. Acts of legislation and deeds of war may confer a high celebrity, but the reputation which they bring is only local and transient; and while they are hailed by the nation which they benefit, they are reprobated by the people whom they ruin or enslave. The labours of science, on the contrary, bear along with them no counterpart of evil. They are the liberal bequests of great minds to every individual of their race, and wherever they are welcomed and honoured, they become the solace of private life, and the ornament and bulwark of the commonwealth.

The importance of Sir Isaac Newton's discoveries has been sufficiently exhibited in the preceding chapters: the peculiar character of his genius, and the method which he pursued in his inquiries, can be gathered only from the study of his works, and from the history of his individual labours. Were we to judge of the qualities of his mind from the early age at which he made his principal discoveries, and from the rapidity of their succession, we should be led to ascribe to him that quickness of penetration, and that exuberance of invention, which is more characteristic of poetical than of philosophical genius. But we must recollect that Newton was placed in the most favourable circumstances for the development of his powers. The flower of his youth, and the vigour of his manhood, were entirely devoted to science. No injudicious guardian controlled his ruling passion, and no ungenial studies or professional toils interrupted the continuity of his pursuits. His discoveries were therefore the fruit of persevering and unbroken study; and he himself declared, that whatever service he had done to the public was not owing to any extraordinary sagacity, but solely to industry and patient thought.

Initiated early into the abstractions of geometry, he was deeply imbued with her cautious spirit; and if his acquisitions were not made with the rapidity of intuition, they were at least firmly secured; and the grasp which he took of his subject was proportional to the mental labour which it had exhausted. Overlooking what was trivial, and separating what was extraneous, he bore down with instinctive sagacity on the prominences of his subject; and having thus grappled with its difficulties, he never failed to entrench himself in its strongholds.

To the highest powers of invention Newton added,

what so seldom accompanies them, the talent of simplifying and communicating his profoundest speculations.* In the economy of her distributions, nature is seldom thus lavish of her intellectual gifts. The inspired genius which creates is rarely conferred along with the matured judgment which combines; and yet without the exertion of both, the fabric of human wisdom could never have been reared. Though a ray from heaven kindled the vestal fire, yet a humble priesthood was required to keep alive the flame.

The method of investigating truth by observation and experiment, so successfully pursued in the *Principia*, has been ascribed by some modern writers of great celebrity to Lord Bacon; and Sir Isaac Newton is represented as having owed all his discoveries to the application of the principles of that distinguished writer. One of the greatest admirers of Lord Bacon has gone so far as to characterize him as a man who has had no rival in the times which are past, and as likely to have none in those which are to come. In a eulogy so overstrained as this, we feel that the language of panegyric has passed into that of idolatry; and we are desirous of weighing the force of arguments which tend to depose Newton from the high priesthood of nature, and to unsettle the proud destinies of Copernicus, Galileo, and Kepler.

That Bacon was a man of powerful genius, and endowed with varied and profound talent,—the most skilful logician,—the most nervous and eloquent writer of the age which he adorned,—are points which have been established by universal suffrage. The study of ancient systems had early impressed him with the conviction that experiment and observation were the only sure guides in physical inquiries; and, ignorant though he was of the methods,

* This valuable faculty characterizes all his writings, whether theological, chemical, or mathematical; but it is peculiarly displayed in his treatise on Universal Arithmetic, and in his Optical Lectures.

the principles, and the details of the mathematical sciences, his ambition prompted him to aim at the construction of an artificial system by which the laws of nature might be investigated, and which might direct the inquiries of philosophers in every future age. The necessity of experimental research, and of advancing gradually from the study of facts to the determination of their cause, though the ground-work of Bacon's method, is a doctrine which was not only inculcated but successfully followed by preceding philosophers. In a letter from Tycho Brahe to Kepler, this industrious astronomer urges his pupil " to lay a solid foundation for his views by actual observation, and then by ascending from these to strive to reach the causes of things;" and it was no doubt under the influence of this advice that Kepler submitted his wildest fancies to the test of observation, and was conducted to his most splendid discoveries. The reasonings of Copernicus, who preceded Bacon by more than a century, were all founded upon the most legitimate induction. Dr. Gilbert had exhibited in his Treatise on the Magnet * the most perfect specimen of physical research. Lionardo da Vinci had described in the clearest manner the proper method of philosophical investigation; † and the whole scientific

* De Magnete, pp. 42, 52, 169, and Pref., p. 80.

† The following passages from Lionardo da Vinci are very striking:—
" Theory is the general and practice the soldiers.

" Experiment is the interpreter of the artifices of nature. It never deceives us; it is our judgment itself which sometimes deceives us, because we expect from it effects which are contrary to experiment. We must consult experiment by varying the circumstances till we have deduced from it general laws; for it is it which furnishes true laws."

" In the study of the sciences which depend on mathematics, those who do not consult nature, but authors, are not the children of nature; they are only her grandchildren. Nature alone is the master of true genius.

" In treating any particular subject, I would first of all make some experiments, because my design is first to refer to experiment, and then

career of Galileo was one continued example of the most
sagacious application of observation and experiment to the
discovery of general laws. The names of Paracelsus,
Van Helmont, and Cardan, have been ranged in opposition
to this constellation of great names, and while it is admitted that even they had thrown off the yoke of the schools,
and had succeeded in experimental research, their cre-
dulity and their pretensions have been adduced as a proof
that to the "bulk of philosophers" the method of induction was unknown. The fault of this argument consists
in the conclusion being infinitely more general than the
fact. The errors of these men were not founded on their
ignorance, but on their presumption. They wanted the
patience of philosophy, and not her methods. An excess
of vanity, a waywardness of fancy, and an insatiable
appetite for that species of passing fame which is derived
from eccentricity of opinion, moulded the reasonings and
disfigured the writings of these ingenious men; and it can
scarcely admit of a doubt, that, had they lived in the present age, their philosophical character would have received
the same impress from the peculiarity of their tempers and
dispositions. This is an experiment, however, which cannot now be made; but the history of modern science supplies the defect, and the experience of every man furnishes
a proof, that in the present age there are many philosophers

to demonstrate why bodies are constrained to act in such a manner. This
is the method which we ought to follow in investigating the phenomena
of nature. It is very true that nature begins by reasoning and ends with
experiment; but it matters not, *we must take the opposite course; as I
have said, we must begin by experiment,* and endeavour by its means to
discover general principles." Thus, says Venturi, spoke Léonard a century before Bacon; and thus, we add, did Leonard tell philosophers all
that they required for the proper investigation of general laws. See
Essai sur les Ouvrages physico-mathematiques de Léonard de Vinci, par
J. B. Venturi. Paris, 1799, pp. 82, 88, &c. See also Carlo Amoretti's
Memoria storiche su la Vita, gli Studi, e le Opere di Lionardo da Vinci.
Milano 1804.

of elevated talents and inventive genius who are as impatient of experimental research as Paracelsus, as fanciful as Cardan, and as presumptuous as Van Helmont.

Having thus shown that the distinguished philosophers who flourished before Bacon were perfect masters both of the principles and practice of inductive research, it becomes interesting to inquire whether or not the philosophers who succeeded him acknowledged any obligation to his system, or derived the slightest advantage from his precepts. If Bacon constructed a method to which modern science owes its existence, we shall find its cultivators grateful for the gift, and offering the richest incense at the shrine of a benefactor whose generous labours conducted them to immortality. No such testimonies, however, are to be found. Nearly two hundred years have gone by, teeming with the richest fruits of human genius, and no grateful disciple has appeared to vindicate the rights of the alleged legislator of science. Even Newton, who was born and educated after the publication of the *Novum Organum*, never mentions the name of Bacon, or his system, and the amiable and indefatigable Boyle treated him with the same disrespectful silence. When we are told, therefore, that Newton owed all his discoveries to the method of Bacon, nothing more can be meant than that he proceeded in that path of observation and experiment which had been so warmly recommended in the *Novum Organum*; but it ought to have been added, that the same method was practised by his predecessors,—that Newton possessed no secret that was not used by Galileo and Copernicus,—and that he would have enriched science with the same splendid discoveries if the name and the writings of Bacon had never been heard of.

From this view of the subject we shall now proceed to examine the Baconian process itself, and consider if it possesses any merit as an artificial method of discovery, or if it is at all capable of being employed, for this purpose, even in the humblest walks of scientific inquiry.

The process of Lord Bacon was, we believe, never tried by any philosopher but himself. As the subject of its application, he selected that of heat. With his usual erudition, he collected all the facts which science could supply,—he arranged them in tables,—he cross-questioned them with all the subtlety of a pleader,—he combined them with all the sagacity of a judge,—and he conjured with them by all the magic of his exclusive processes. But, after all this display of physical logic, nature thus interrogated was still silent. The oracle which he had himself established refused to give its responses, and the ministering priest was driven with discomfiture from his own shrine. This example, in short, of the application of his system, will remain to future ages as a memorable instance of the absurdity of attempting to fetter discovery by any artificial rules.

Nothing even in mathematical science can be more certain than that a collection of scientific facts are of themselves incapable of leading to discovery, or to the determination of general laws, unless they contain the predominating fact or relation in which the discovery mainly resides. A vertical column of arch-stones possesses more strength than the same materials arranged in an arch without the key-stone. However nicely they are adjusted, and however nobly the arch may spring, it never can possess either equilibrium or stability. In this comparison all the facts are supposed to be necessary to the final result; but, in the inductive method, it is impossible to ascertain the relative importance of any facts, or even to determine if the facts have any value at all, till the master fact which constitutes the discovery has crowned the zealous efforts of the aspiring philosopher. The mind then returns to the dark and barren waste over which it has been hovering; and by the guidance of this single torch it embraces, under the comprehensive grasp of general principles, the multifarious and insulated phenomena which had formerly neither value nor connexion. Hence it must be obvious to the most superficial thinker, that discovery consists either in the detection of some concealed relation, some deep-seated affinity which baffles ordinary research, or in the discovery of some simple fact which is connected by slender ramifications with the subject to be investigated; but which, when once detected, carries us back by its divergence to all the phenomena which it embraces and explains.

In order to give additional support to these views, it would be interesting to ascertain the general character of the process by which a mind of acknowledged power actually proceeds in the path of successful inquiry. The history of science does not furnish us with much information on this head; and if it is to be found at all, it must be gleaned from the biographies of eminent men. Whatever this process may be in its details, if it has any, there cannot be the slightest doubt that in its generalities at least it is the very reverse of the method of induction. The impatience of genius spurns the restraints of mechanical rules, and never will submit to the plodding drudgery of inductive discipline. The discovery of a new fact unfits even a patient mind for deliberate inquiry. Conscious of having added to science what had escaped the sagacity of former ages, the ambitious spirit invests its new acquisition with an importance which does not belong to it. He imagines a thousand consequences to flow from his discovery: he forms innumerable theories to explain it; and he exhausts his fancy in trying all its possible relations to recognized difficulties and unexplained facts. The reins, however, thus freely given to his imagination, are speedily drawn up. His wildest conceptions are all subjected to

the regid test of experiment, and he has thus been hurried
by the excursions of his own fancy into new and fertile
paths, far removed from ordinary observation.   Here the
peculiar character of his own genius displays itself by the
invention of methods of trying his own speculations, and
he is thus often led to new discoveries far more important
and general than that by which he began his inquiry.
For a confirmation of these views, we may refer to the
History of Kepler's Discoveries; and if we do not recog-
nize them to the same extent in the labours of Newton, it
is because he kept back his discoveries till they were nearly
perfected, and therefore withheld the successive steps of
his inquiries.

The social character of Sir Isaac Newton was such as
might have been expected from his intellectual attainments.   He was modest, candid, and affable, and without
any of the eccentricities of genius, suiting himself to every
company, and speaking of himself and others in such a
manner that he was never even suspected of vanity.   "But
this," says Dr. Pemberton, "I immediately discovered in
him, which at once both surprised and charmed me.
Neither his extreme great age, nor his universal reputation, had rendered him stiff in opinion, or in any degree
elated.   Of this I had occasion to have almost daily experience.   The remarks I continually sent him by letters
on the *Principia* were received with the utmost goodness.
These were so far from being any ways displeasing to him,
that on the contrary it occasioned him to speak many
kind things of me to my friends, and to honour me with
a public testimony of his good opinion."

The modesty of Sir Isaac Newton, in reference to his
great discoveries, was not founded on any indifference to
the fame which they conferred, or upon any erroneous
judgment of their importance to science.   The whole of
his life proves. that he knew his place as a philosopher,

and was determined to assert and vindicate his rights.
His modesty arose from the depth and extent of his knowledge, which showed him what a small portion of nature
he had been able to examine, and how much remained to
be explored in the same field in which he had himself
laboured. In the magnitude of the comparison he recognized his own littleness; and a short time before his death
he uttered this memorable sentiment: "I do not know
what I may appear to the world; but to myself I seem to
have been only like a boy playing on the sea-shore, and
diverting myself in now and then finding a smoother pebble
or a prettier shell than ordinary, whilst the great ocean of
truth lay all undiscovered before me." What a lesson to
the vanity and presumption of philosophers,—to those
especially who have never even found the smoother pebble
or the prettier shell ! What a preparation for the latest
inquiries and the last views of the decaying spirit,—for
those inspired doctrines which alone can throw a light
over the dark ocean of undiscovered truth !

The native simplicity of Sir Isaac Newton's mind is
finely pourtrayed in the affecting letter in which he
acknowledges to Locke, that he had thought and spoken
of him uncharitably; and the humility and candour in
which he asks forgiveness, could have emanated only from
a mind as noble as it was pure.

In the religious and moral character of our author,
there is much to admire and to imitate. While he exhibited in his life and writings an ardent regard for the
general interests of religion, he was at the same time a
firm believer in Revelation. He was too deeply versed in
the Scriptures, and too much imbued with their spirit, to
judge harshly of other men who took different views of
them from himself. He cherished the great principles of
religious toleration, and never scrupled to express his
abhorrence of persecution, even in its mildest form.

Immorality and impiety he never permitted to pass unre-
proved; and when Dr. Halley * ventured to say any
thing that appeared disrespectful to religion, he invariably
checked him, with the remark, "I have studied these
things,—you have not." †

After Sir Isaac Newton took up his residence in London, he lived in a very handsome style, and kept his
carriage, with an establishment of three male and three
female servants. In his own house he was hospitable and
kind, and on proper occasions he gave splendid entertain-
ments, though without ostentation or vanity. His own
diet was frugal, and his dress was always simple; but on
one occasion, when he opposed the Hon. Mr. Annesley in
1705, as a candidate for the university, he is said to have
put on a suit of laced clothes.

His generosity and charity had no bounds, and he used
to remark, that they who gave away nothing till they died,
never gave at all. Though his wealth had become considerable by a prudent economy, yet he had always a
contempt for money, and he spent a considerable part of
his income in relieving the poor, in assisting his relations,
and in encouraging ingenuity and learning. The sums

Mr. Hearne, in a memorandum dated April 4th, 1726, states, that a
great quarrel happened between Sir Isaac Newton and Mr. Halley. If this
is true, the difference is likely to have originated from this cause.

† This anecdote was related by Professor Rigaud of Oxford, on the
authority of Dr. Maskelyne. It is only right to add here that Brewster,
in his enlarged edition, gives reasons for believing that what led to Newton's rebuke of Halley on religious grounds, "amounted to nothing more
than his maintaining certain opinions about the existence of a pre-Adamite
earth, and ridiculing vulgar errors which have been too frequently associated with religious truth." He refers us to an able *Defence of Halley
against the Charge of Religious Infidelity*, by the Rev. S. J. Rigaud (son
of Professor Rigaud), published at Oxford in 1844. Flamsteed, indeed,
never scrupled to impute such views to him, and to bring charges also
against his character; but he appears to have been apt to do this without
sufficient cause.—EDITOR.

which he gave to his relations at different times were enormous;* and in 1724 he wrote a letter to the Lord Provost of Edinburgh, offering to contribute £20 per annum to a provision for Mr. Maclaurin, provided he accepted the situation of assistant to Mr. James Gregory, who was Professor of Mathematics in the University.

The habits of deep meditation which Sir Isaac Newton had acquired, though they did not show themselves in his intercourse with society, exercised their full influence over his mind when in the midst of his own relations. Absorbed in thought, he would often sit down on his bedside after he rose, and remain there for hours without dressing himself, occupied with some interesting investigation which had fixed his attention. Owing to the same absence of mind, he neglected to take the requisite quantity of nourishment, and it was therefore often necessary to remind him of his meals.†

Sir Isaac Newton is supposed to have had little knowledge of the world, and to have been very ignorant of the habits of society. This opinion has, we think, been rashly

* "He was very kind to all the Ayscoughs. To one he gave £800, to another £200, and to a third £100, and many other sums; and other engagements did he enter into also for them. He was the ready assistant of all who were any way related to him,—to their children and grandchildren."—*Annual Register*, 1776, vol. xix., p. 25. It is on record also, that Sir Isaac gave some donations to the parish of Colsterworth, in which he was born.

† The following anecdote of Sir Isaac's absence has been published, but I cannot vouch for its authenticity. His intimate friend, Dr. Stukely, who had been deputy to Dr. Halley as secretary to the Royal Society, was one day shown into Sir Isaac's dining-room, where his dinner had been for some time served up. Dr. Stukely waited for a considerable time; and, getting impatient, he removed the cover from a chicken, which he ate, replacing the bones under the cover. In a short time Sir Isaac entered the room, and after the usual compliments sat down to his dinner, but, on taking off the cover, and seeing nothing but the bones, he remarked, "How absent we philosophers are! I really thought I had not dined."

deduced from a letter which he wrote in the twenty-seventh year of his age to his young friend, Francis Aston, Esq., who was about to set out on his travels. This letter is a highly interesting production; and while it shows much knowledge of the human heart, it throws a strong light upon the character and opinions of its author.

In his personal appearance, Sir Isaac Newton was not above the middle size, and in the latter part of his life was inclined to be corpulent. According to Mr. Conduitt,[*] "he had a very lively and piercing eye, a comely and gracious aspect, with a fine head of hair as white as silver, without any baldness, and, when his peruke was off, was a venerable sight." Bishop Atterbury asserts,[†] on the other hand, that the lively and piercing eye did not belong to Sir Isaac during the last twenty years of his life. "Indeed," says he, "in the whole air of his face and make there was nothing of that penetrating sagacity which appears in his compositions. He had something rather languid in his look and manner which did not raise any great expectation in those that did not know him." This opinion of Bishop Atterbury is confirmed by an observation of Mr. Thomas Hearne,[‡] who says, that "Sir Isaac was a man of no very promising aspect. He was a short well-set man. He was full of thought, and spoke very little in company, so that his conversation was not agreeable. When he rode in his coach, one arm would be out of his coach on one side and the other on the other." Sir Isaac never wore spectacles, and never "lost more than one tooth to the day of his death."

---

* At this final reference to Mr. Conduitt, it is well to mention that the above appears to be the correct way of spelling his name, though in the first chapter we inadvertently left it as Brewster did there in both editions, "Conduit."—EDITOR.

† Epistolary Correspondence, vol. I, p. 160, sect. 77.

‡ MS. Memoranda in the Bodleian Library.

Besides the statue of Sir Isaac Newton executed by Roubiliac, there is a bust of him by the same artist in the Library of Trinity College, Cambridge. Several good paintings of him are extant. Two of these are in the hall of the Royal Society of London, and have, we believe, been often engraved. Another, by Vanderbank, is in the small combination room in Trinity College, and has been engraved by Vertue. Another, by Valentine Ritts, is in the Hall of Trinity College; but the best, from which our engraving is copied, was painted by Sir Godfrey Kneller, and is in the possession of Lord Egremont at Petworth. In the University Library there is preserved a cast taken from his face after death.

Every memorial of so great a man as Sir Isaac Newton has been preserved and cherished with peculiar veneration. His house at Woolsthorpe, of which we have given an engraving, has been religiously protected by Mr. Turnor of Stoke Rocheford, the proprietor. Dr. Stukely, who visited it in Sir Isaac's lifetime on the 13th of October, 1721, gives the following description of it in his letter to Dr. Mead, written in 1727: "'Tis built of stone, as is the way of the country hereabouts, and a reasonable good one. They led me up stairs and showed me Sir Isaac's study, where I suppose he studied when in the country in his younger days, or perhaps when he visited his mother from the University. I observed the shelves were of his own making, being pieces of deal boxes which probably he sent his books and clothes down in on those occasions. There were some years ago two or three hundred books in it of his father-in-law, Mr. Smith, which Sir Isaac gave to Dr. Newton of our town."*

When the house was repaired in 1798, a tablet of white marble was put up by Mr. Turnor in the room where Sir Isaac was born, with the following inscription :—

* Turnor's Collections, p. 176.

"Sir Isaac Newton, son of John Newton, Lord of the Manor of Woolsthorpe, was born in this room on the 25th of December, 1642.

> "Nature and Nature's laws lay hid in night ;
> God said, ' Let Newton be,' and all was Light."

The following lines have been written upon the house :—

> "Here Newton dawned, here lofty wisdom woke,
> And to a wondering world divinely spoke.
> If Tully glowed, when Phædrus' steps he trode,
> Or fancy formed Philosophy a God ;
> If sages still for Homer's birth contend,
> The Sons of Science at this dome must bend.
> All hail the shrine! All hail the natal day!
> Cam boasts his noon,—This *Cot* his morning ray."

The house * is now occupied by a farmer of the name of John Woollerton. It still contains the two dials made by Newton, but the styles of both are wanting. The celebrated apple tree, the fall of one of the apples of which is said to have turned the attention of Newton to the subject of gravity, was destroyed by wind about four years ago; but Mr. Turnor has preserved it in the form of a chair.

The chambers which Sir Isaac inhabited at Cambridge are known by tradition. They are the apartments next to the great gate of Trinity College, and it is believed that they then communicated by a staircase with the Observatory in the Great Tower,—an observatory which was furnished by the contributions of Newton, Cotes, and others. His telescope, represented in Fig. 3, page 29, is

---

* We have already, in a note to the first chapter, referred to our visit there in the autumn of 1878. On approaching Woolsthorpe, after leaving Colsterworth, we asked a country-man which was Mr. Woollerton's? The reply was, "That house round the corner, Sir, is Sir Isaac Newton's," with an air of "That's what you want, I am sure." A very pleasant walk afterwards through North and South Witham led us to Market Overton, the birth-place of Newton's mother, and afterwards down to Oakham, the county-town of Rutland, in the vale of Catmose.—EDITOR.

preserved in the library of the Royal Society of London, and his globe, his universal ring-dial, quadrant, compass, and a reflecting telescope said to have belonged to him, in the library of Trinity College. There is also in the same collection three locks of his silver-white hair. The door of his bookcase is in the Museum of the Royal Society of Edinburgh.

The manuscripts, letters, and other papers of Newton have been preserved in different collections. His correspondence with Cotes relative to the second edition of the Principia, and amounting to between sixty and a hundred letters, a considerable portion of the manuscript of that work, and five letters to Dr. Keill on the Leibnitzian controversy, are preserved in the library of Trinity College, Cambridge.* Newton's letters to Flamsteed, about thirty-four in number, are deposited in the library of Corpus Christi College, Oxford, and those of Flamsteed are at Hurstbourne Park. In the British Museum, and in the Royal Society of London, there are many letters of Newton and his correspondents. Several letters of Sir Isaac, and the original specimen which he drew up of the Principia, exist among the papers of Mr. William Jones (the father of Sir William Jones), which are preserved at Shirburn Castle, in the library of Lord Macclesfield. But the great mass of Newton's papers came into possession of the Portsmouth family through his niece, Lady Lymington, and have been safely preserved by that noble family.

* All these letters, with others, particularly several to Oldenburg, were published in 1850 by Mr. Edleston, Fellow of Trinity College, at the expense of the Master and Seniors of the College, under the title "Correspondence of Sir Isaac Newton and Professor Cotes, &c."—EDITOR.

APPENDIX.

—

No. 1.

OBSERVATIONS ON THE FAMILY OF SIR ISAAC NEWTON.

IN the year 1705, Sir Isaac gave into the Herald's Office an elaborate pedigree, stating, upon oath, *that he had reason to believe* that John Newton of Westby, in the county of Lincoln, was his great-grandfather's father, and that this was the same John Newton who was buried in Basingthorpe Church on the 22nd of December 1563. This John Newton had four sons, John, Thomas, Richard, and William Newton of Gunnerly, the last of whom was great-grandfather to Sir John Newton, Bart., of Hather. Sir Isaac considered himself as descended from the eldest of these, *he having, by tradition from his kindred ever since he can remember, reckoned himself next of kin (among the Newtons) to Sir John Newton's family.*

The pedigree, founded upon these and other considerations, was accompanied by a certificate from Sir John Newton, of Thorpe, Bart., who states that he had heard his father speak of Sir Isaac Newton *as of his relation and kinsman,* and that *he himself believed that Sir Isaac was descended from John Newton, son to John Newton of Westby, but knoweth not in what particular manner.*

The pedigree of Sir Isaac, as entered at the Herald's Office, does not seem to have been satisfactory either to himself or to his successors, as it could not be traced with certainty beyond his grandfather; and it will be seen from the following interesting correspondence, that, upon making farther researches, he had found some reason to believe that he was of Scotch extraction.

Extract of a Letter from the Rev. Dr. Reid of Glasgow to Dr. Gregory of Edinburgh, dated 14th of March, 1784.

"I SEND you on the other page an anecdote respecting Sir Isaac Newton, which I do not remember whether I ever happened to mention to you in conversation. If his descent be not clearly ascertained, (as I think it is not in the books I have seen,) might it not be worth while to inquire if evidence can be found to confirm the account which he is said to have given of himself? Sheriff Cross was very zealous about it when death put a stop to his inquiries.

"When I lived in Old Aberdeen above twenty years ago, I happened to be conversing over a pipe of tobacco with a gentleman of that country, who had been lately at Edinburgh. He told me that he had been often in company with Mr. Hepburn of Keith, with whom I had the honour of some acquaintance. He said that, speaking of Sir Isaac Newton, Mr. Hepburn mentioned an anecdote, which he had from Mr. James Gregory, Professor of Mathematics at Edinburgh, which was to this purpose:—

"Mr. Gregory being at London for some time after he resigned the mathematical chair, was often with Sir Isaac Newton. One day Sir Isaac said to him, 'Gregory, I believe you don't know that I am connected with Scotland.'—'Pray how, Sir Isaac?' said Gregory. Sir Isaac said he was told that his grandfather was a gentleman of East Lothian; that he came to London with King James at his accession to the crown of England, and there spent his fortune, as many more did at that time, by which his son (Sir Isaac's father) was reduced to mean circumstances. To this Gregory bluntly replied, 'Newton a gentleman of East Lothian? I never heard of a gentleman of East

Lothian of that name.'   Upon this Sir Isaac said, 'that being very young when his father died, he had it only by tradition, and it might be a mistake,' and immediately turned the conversation to another subject.

"I confess I suspected that the gentleman who was my author had given some colouring to this story, and therefore I never mentioned it for a good many years.

"After I removed to Glasgow, I came to be very intimately acquainted with Mr. Cross, then Sheriff of Lanark, and one day at his own house mentioned this story, without naming my author, of whom I expressed some diffidence.

"The Sheriff immediately took it up as a matter worth being inquired into.   He said he was well acquainted with Mr. Hepburn of Keith, (who was then alive,) and that he would write him to know whether he ever heard Mr. Gregory say that he had such a conversation with Sir Isaac Newton.   He said he knew that Mr. Keith, the ambassador, was also intimate with Mr. Gregory, and that he would write him to the same purpose.

"Some time after, Mr. Cross told me that he had answers from both the gentlemen above mentioned, and that both remembered to have heard Mr. Gregory mention the conversation between him and Sir Isaac Newton, to the purpose above narrated, and at the same time acknowledged that they had made no farther inquiry about the matter.

"Mr. Cross, however, continued the inquiry; and, a short time before his death, told me that all he had learned was, that there is, or was lately, a baronet's family of the name of Newton in West Lothian or Mid-Lothian, (I have forgot which:) that there is a tradition in that family, that Sir Isaac Newton wrote a letter to the old knight that then was, (I think Sir John Newton of Newton was his name,) desiring to know what children, and particularly what sons he had, their age, and what professions they intended: that the old baronet never deigned to return an answer to this letter, which his family was sorry for, as they thought Sir Isaac might have intended to do something for them."

Several years after this letter was written, a Mr. Barron, a relation of Sir Isaac Newton, seems to have been making inquiries respecting the family of his ancestor, and in consequence of this the late Professor Robison applied to Dr. Reid, to obtain from him a more particular account of the remarkable conversation between Sir Isaac and Mr. James Gregory, referred to in the preceding letter. In answer to this request, Dr. Reid wrote the following letter, for which I am indebted to John Robison, Esq., Sec. R.S.E., who found it among his father's manuscripts.

Letter from Dr. Reid to Professor Robison respecting the Family of Sir Isaac Newton.

" DEAR SIR,

" I am very glad to learn by your's of April 4th, that a Mr. Barron, a near relation of Sir Isaac Newton, is anxious to inquire into the descent of that great man, as the family cannot trace it further, with any certainty, than his grandfather. I therefore, as you desire, send you a precise account of all I know; and am glad to have this opportunity, before I die, of putting this information in hands that will make the proper use of it, if it shall be found of any use.

" Several years before I left Aberdeen, (which I did in 1764,) Mr. Douglas of Fechel, the father of Sylvester Douglas, now a barrister at London, told me that, having been lately at Edinburgh, he was often in company with Mr. Hepburn of Keith, a gentleman of whom I had some acquaintance, by his lodging a night at my house, at New Machar, when he was in the rebel army in 1745. That Mr. Hepburn told him, that he had heard Mr. James Gregory,

Professor of Mathematics, Edinburgh, say that, being one day in familiar conversation with Sir Isaac Newton, at London, Sir Isaac said, 'Gregory, I believe you don't know that I am a Scotchman.'—'Pray how is that?' said Gregory. Sir Isaac said he was informed that his grand-father (or great-grandfather) was a gentleman of East (or West) Lothian: that he went to London with King James the First at his accession to the crown of England: and that he attended the Court in expectation, as many others did, until he spent his fortune, by which means his family was reduced to low circumstances. At the time this was told me, Mr. Gregory was dead, otherwise I should have had his own testimony, for he was my mother's brother. I likewise thought at that time, that it had been certainly known that Sir Isaac had been descended from an old English family, as I think is said in his *éloge* before the Academy of Sciences at Paris, and therefore I never mentioned what I had heard for many years, believing that there must be some mistake in it.

"Some years after I came to Glasgow, I mentioned (I believe for the first time) what I had heard to have been said by Mr. Hepburn, to Mr. Cross, late sheriff of this county, whom you will remember. Mr. Cross was moved by this account, and immediately said, 'I know Mr. Hepburn very well, and I know he was intimate with Mr. Gregory; I shall write him this same night, to know whether he heard Mr. Gregory say so, or not.' After some reflection, he added, 'I know that Mr. Keith, the ambassador, was also an intimate acquaintance of Mr. Gregory, and as he is at present in Edinburgh, I shall likewise write to him this night.'

"The next time I waited on Mr. Cross, he told me that he had wrote both to Mr. Hepburn and Mr. Keith, and had an answer from both, and that both of them testified that they had several times heard Mr. James Gregory say, that Sir Isaac Newton told him what is above expressed, but that neither they, nor Mr. Gregory, as far as they knew, ever made any farther inquiry into the matter. This appeared very strange, both to Mr. Cross and me, and he said he would reproach them for their indifference, and would make inquiry as soon as he was able.

"He lived but a short time after this; and, in the last conversation I had with him upon the subject, he said, that all he had yet learned was, that there was a Sir John Newton of Newton, in one of the counties of Lothian, (but I have forgot which,) some of whose children were yet alive; that they reported that their father, Sir John, had a letter from Sir Isaac Newton, desiring to know the state of his family, what children he had, particularly what sons, and in what way they were. The old knight never returned an answer to this letter, thinking probably that Sir Isaac was some upstart who wanted to claim a relation to his worshipful house. This omission the children regretted, conceiving that Sir Isaac might have had a view of doing something for their benefit.

"After this I mentioned occasionally in conversation what I knew, hoping that these facts might lead to some more certain discovery, but I found more coldness about the matter than I thought it deserved. I wrote an account of it to Dr. Gregory, your colleague, that he might impart it to any member of the Antiquarian Society, who he judged might have the curiosity to trace the matter farther.

"In the year 1787, my colleague, Mr. Patrick Wilson, Professor of astronomy, having been in London, told me on his return that he had met accidentally with a James Hutton, Esq., of Pimlico, Westminster, a near relation of Sir Isaac Newton,* to whom he mentioned what he had heard from me with respect to Sir Isaac's descent, and that

---

* See page 290, Note.

I wished much to know something more decisive on that
subject.   Mr. Hutton said, if I pleased to write to him he
would give me all the information he could give.   I wrote
him accordingly, and had a very polite answer, dated at
Bath, 25th of December, 1787, which is now before me.
He says, 'I shall be glad, when I return to London, if I
can find, in some old notes of my mother, any thing that
may fix the certainty of Sir Isaac's descent.   If he spoke
so to Mr. James Gregory, it is most certain he spoke truth.
But Sir Isaac's grandfather, not his great-grandfather,
must be the person who came from Scotland with King
James I.   If I find any thing to the purpose, I will take
care it shall reach you.'

"In consequence of this letter I expected another from
Mr. Hutton when he should return to London, but have
never had any.   Mr. Wilson told me he was a very old
man, and whether he be dead or alive I know not.

"This is all I know of the matter, and, for the facts above
mentioned, I pledge my veracity.   I am much obliged
to you, dear Sir, for the kind expression of your affection
and esteem, which, I assure you, are mutual on my part,
and I sincerely sympathize with you on your afflicting
state of health, which makes you consider yourself as out
of the world, and despair of seeing me any more.

"I have been long out of the world by deafness and
extreme old age.   I hope, however, if we should not meet
again in this world, that we shall meet and renew our
acquaintance in another.   In the mean time, I am, with
great esteem, dear Sir, yours affectionately,

"*Glasgow College,*   THOS. REID,"
    12*th April*, 1792."

This curious letter I published in the Ed. Phil. Journal
for October, 1820.   It excited the particular attention of
the late George Chalmers, Esq., who sent me an elaborate

letter upon the subject; but as I was at that time in the
expectation of obtaining some important information
through other channels, this letter was not published.
This hope, however, has been disappointed. A careful
search has been made through the charter-chest of the
Newtons of Newton in East Lothian, by Mr. Richard Hay
Newton, the representative of that family, but no document whatever has been found that can throw the least
light upon the matter. It deserves to be remarked, however, that Sir Richard Newton, the alleged correspondent
of Sir Isaac, appears to have destroyed his correspondence;
for though the charter-chest contained the letters of his
predecessors for some generations, yet there is not a single
epistolary document either of his own or of his lady's.

Hitherto the evidence of Sir Isaac's Scottish descent has
been derived chiefly from his conversation with Mr. James
Gregory; but I am enabled, by the kindness of Mr. Robison, to corroborate this evidence by the following information, derived, as will be seen, from the family of the
Newtons, of Newton. Among various memoranda in the
handwriting of Professor Robison, who at one time proposed to write the life of Sir Isaac, are the following:—

" 1st. Lord Henderland informed me in a letter dated
March, 1794, that he had heard from his infancy that
Sir Isaac considered himself as descended from the family
of Newton, of Newton. This he heard from his uncle
Richard Newton, of Newton, (who was third son of Lord
William Hay, of Newhall.) He said that Sir Isaac wrote
to Scotland to learn whether any descendants of that
family remained, and this (it was thought) with the view
to leave some of his fortune to the family possessing the
estate with the title of baronet. Mr. Newton not having
this honour, and being a shy man, did not encourage the
correspondence, because he did not consider *himself* as of
kin to Sir Isaac," &c.

"2nd. Information communicated to me by Hay Newton, Esq., of that ilk, 18th of August, 1800."

"The late Sir Richard Newton, of Newton, Bart., chief of that name, having no male children, settled the estate and barony of Newton, in East Lothian county, upon his relation, Richard Hay Newton, Esq., son of Lord William Hay." *—"It cannot be discovered how long the family of Newton have been in possession of the barony, there being no tradition concerning that circumstance further than that they came originally from England at a very distant period. and settled on these lands."—"The celebrated Sir Isaac Newton was a distant relation of the family, and corresponded with the last baronet, the above-mentioned Sir Richard Newton."

The preceding documents furnish the most complete evidence that the conversation respecting Sir Isaac's family took place between him and Mr. Gregory; and the testimony of Lord Henderland proves that his own uncle, Richard Newton, of Newton, the immediate successor of Sir Richard Newton, with whom Sir Isaac corresponded, was perfectly confident that such a correspondence took place.

All these circumstances prove that Sir Isaac Newton could not trace his pedigree with any certainty beyond his grandfather, and that there were two different traditions in his family, one which referred his descent to John Newton of Westby, and the other to a gentleman of East Lothian who accompanied King James VI. to England. In the first of these traditions he seems to have placed most confidence in 1705, when he drew out his traditionary pedigree, but as the conversation with Professor James Gregory respecting his Scotch extraction took place *twenty years* afterwards, namely, between 1725 and 1727,

---

* This entail was executed in 1724, a year or two before Sir Richard's death.

it is probable that he had discovered the incorrectness of
his first opinions, or at least was disposed to attach more
importance to the other tradition, respecting his descent
from a Scotch family.

In the letter addressed to me by the learned George
Chalmers, Esq., I find the following observations respecting the immediate relations of Sir Isaac. "The Newtons,
of Woolsthorpe," says he, "who were merely yeomen
farmers, were not by any means opulent. The son of
Sir Isaac's father's brother was a carpenter called John.
He was afterwards appointed gamekeeper to Sir Isaac,
as lord of the manor, and died at the age of sixty in 1725.
This John had a son, Robert, (John ?) who was Sir Isaac's
second cousin, and who became possessed of the whole land
estates at and near Woolsthorpe, which belonged to the great
Newton, as his heir-at-law.* Robert (John ?) became a
worthless and dissolute person, who very soon wasted this
ancient patrimony; and, falling down with a tobacco-pipe
in his mouth when he was drunk, it broke in his throat,
and put an end to his life at the age of thirty years, in
1737."

No. II.

LETTER FROM SIR ISAAC NEWTON TO FRANCIS ASTON,
Esq., A YOUNG FRIEND WHO WAS ON THE EVE OF
SETTING OUT UPON HIS TRAVELS.

MR. ASTON was elected a Fellow of the Royal Society,
in 1678. He held the office of Secretary between 1681
and 1685; and he was the author of some observations on
certain unknown ancient characters, which were published
in the Philosophical Transactions for 1693.

* See p. 293.

The letter in question (which is referred to in page 306)
was written when Newton was only twenty-six years of
age.   It is in every respect an interesting document :—

" Trinity College, Cambridge,

" Sir,   May 18th, 1669.

" Since in your letter you give mee so much liberty of
spending my judgement about what may be to your
advantage in travelling, I shall do it more freely than perhaps otherwise would have been decent.   First, then, I
will lay down some general rules, most of which, I believe,
you have considered already ; but, if any of them be new
to you, they may excuse the rest ; if none at all, yet is my
punishment more in writing than your's in reading.

" When you come into any fresh company, 1. Observe
their humours.   2. Suit your .own carriage thereto, by
which insinuation you will make their converse more free
and open.   3. Let your discourse be more in query's and
doubtings than peremptory assertions or disputings, it
being the designe of travellers to learne, not to teach.
Besides, it will persuade your acquaintance that you have
the greater esteem of them, and soe make them more
ready to communicate what they know to you ; whereas
nothing sooner occasions disrespect and quarrels than
peremptorinesse.   You will find little or no advantage in
seeming wiser, or much more ignorant, than your company.
4. Seldom discommend any thing, though never so bad,
or doe it but moderately, lest you bee unexpectedly forced
to an unhansom retraction.   It is safer to commend any
thing more than it deserves, than to discommend a thing
soe much as it deserves ; for commendations meet not soe
often with oppositions, or, at least, are not usually soe ill
resented by men that think otherwise, as discommendations ;
and you will insinuate into men's favour by nothing sooner
than seeming to approve and commend what they like ;

but beware of doing it by a comparison. 5. If you be affronted, it is better, in a forraine country, to pass it by in silence, and with a jest, though with some dishonour, than to endeavour revenge; for, in the first case, your credit's ne'er the worse when you returu into England, or come into other company that have not heard of the quarrell. But, in the second case, you may beare the marks of the quarrell while you live, if you outlive it at all. But, if you find yourself unavoidably engaged, 'tis best, I think, if you can command your passion and language, to keep them pretty evenly at some certain moderate pitch, not much hightning them to exasperate your adversary, or provoke his friends, nor letting them grow over much dejected to make him insult. In a word, if you can keep reason above passion, that and watchfullnesse will be your best defeudants. To which purpose you may consider, that, though such excuses as this,—He provok't mee so much I could not forbear,—may pass among friends, yet amongst strangers they are insignificant, and only argue a traveller's weaknesse.

"To these I may add some general heads for inquiry's or observations, such as at present I can think on. As, 1. To observe the policys, wealth, and state-affairs of nations, so far as a solitary traveller may conveniently doe. 2. Their impositions upon all sorts of people, trades, or commoditys, that are remarkable. 3. Their laws and customs, how far they differ from ours. 4. Their trades and arts, wherein they excell or come short of us in England. 5. Such fortifications as you shall meet with, their fashion, strength, and advantages for defence, and other such military affairs as are considerable. 6. The power and respect belonging to their degrees of nobility or magistracy. 7. It will not be time mispent to make a catalogue of the names and excellencys of those men that are most wise, learned, or esteemed in any nation. 8. Observe the me-

chanism and manner of guiding ships. 9. Observe the
products of nature in several places, especially in mines,
with the circumstances of mining and of extracting metals
or minerals out of their oare, and of refining them; and
if you meet with any transmutations out of their own
species into another, (as out of iron into copper, out of
any metall into quicksilver, out of one salt into another,
or into an insipid body, &c.,) those, above all, will be
worth your noting, being the most luciferous, and many
times lucriferous experiments too in philosophy. 10. The
prices of diet and other things. 11. And the staple commoditys of places.

"These generals, (such as at present I could think of,)
if they will serve for nothing else, yet they may assist you
in drawing up a modell to regulate your travells by. As
for particulars, these that follow are all that I can now
think of, viz., Whether at Schemnitium, in Hungary,
(where there are mines of gold, copper, iron, vitriol, anti-
mony, &c.,) they change iron into copper by dissolving it
in a vitriolate water, which they find in cavitys of rocks in
the mines, and then melting the slimy solution in a strong
fire, which in the cooling proves copper. The like is said
to be done in other places, which I cannot now remember;
perhaps, too, it may be done in Italy. For about twenty
or thirty years agone there was a certain vitrioll came from
thence, (called Roman vitrioll,) but of a nobler virtue
than that which is now called by that name; which vitrioll
is not now to be gotten, because, perhaps, they make a
greater gain by some such trick as turning iron into copper with it, than by selling it. 2. Whether, in Hungary,
Sclavonia, Bohemia, near the town Eila, or at the mountains of Bohemia near Silesia, there be rivers whose waters
are impregnated with gold; perhaps the gold being dissolved by some corrosive waters like *aqua regis*, and the
solution carried along with the streame, that runs through

the mines.   And whether the practice of laying mercury in the river, till it be tinged with gold, and then straining the mercury through leather, that the gold may stay behind, be a secret yet, or openly practised.   3. There is newly contrived, in Holland, a mill to grind glasses plane withall, and I think polishing them too; perhaps it will be worth the while to see it.   4. There is in Holland one —— Borry, who some years since was imprisoned by the Pope, to have extorted from him secrets (as I am told) of great worth, both as to medicine and profit, but he escaped into Holland, where they have granted him a guard.   I think he usually goes cloathed in green.   Pray inquire what you can of him, and whether his ingenuity be any profit to the Dutch.   You may inform yourself whether the Dutch have any tricks to keep their ships from being all worm-eaten in their voyages to the Indies.   Whether pendulum clocks do any service in finding out the longitude, &c.

"I am very weary, and shall not stay to part with a long compliment, only I wish you a good journey, and God be with you.

"Is. NEWTON."

"Pray let us hear from you in your travells.   I have given your two books to Dr. Arrowsmith."

No. III.

"A REMARKABLE AND CURIOUS CONVERSATION BETWEEN Sir Isaac Newton and Mr. Conduitt.

"I WAS on Sunday night the 7th of March, 1724–5, at Kensington with Sir Isaac Newton, in his lodgings, just after he was come out of a fit of the gout, which he had had in both his feet, for the first time, in the eighty-third

year of his age. He was better after it, and his head
clearer, and memory stronger, than I had known them for
some time.   He then repeated to me, by way of discourse
very distinctly, though rather in answer to my queries than
in one continued narration, what he had often hinted to
me before, viz., that it was his conjecture, (he would
affirm nothing,) that there was a sort of revolution in the
heavenly bodies; that the vapours and light emitted by
the sun, which had their sediment as water, and other
matter, had gathered themselves by degrees into a body,
and attracted more matter from the planets, and at last
made a secondary planet, (viz., one of those that go round
another planet,) and then by gathering to them, and
attracting more matter, became a primary planet; and then
by increasing still became a comet, which, after certain
revolutions, by coming nearer and nearer to the sun, had
all its volatile parts condensed, and became a matter fit to
recruit and replenish the sun (which must waste by the
constant heat and light it emitted) as a faggot would this
fire if put into it, (we were sitting by a wood fire,) and
that that would probably be the effect of the comet of 1680,
sooner or later; for, by the observations made upon it, it
appeared, before it came near the sun, with a tail only
two or three degrees long; but by the heat it contracted
in going so near the sun, it seemed to have a tail of thirty
or forty degrees when it went from it; that he could not
say when this comet would drop into the sun; it might,
perhaps, have five or six revolutions more first, but whenever it did, it would so much increase the heat of the sun
that this earth would be burnt, and no animals in it could
live.   That he took the three phenomena seen by Hipparchus, Tycho Brahe, and Kepler's disciples, to have been
of this kind; for he could not otherwise account for an
extraordinary light, as those were, appearing all at once
among the fixed stars (all which he took to be suns en-

lightening other planets as our sun does ours) as big as
Mercury or Venus seems to us, and gradually diminishing
for sixteen months, and then sinking into nothing. He
seemed to doubt whether there were not intelligent beings
superior to us who superintended these revolutions of the
heavenly bodies by the direction of the Supreme Being.
He appeared also to be very clearly of opinion that the
inhabitants of this world were of a short date, and alleged
as one reason for that opinion, that all arts, as letters,
ships, printing, needle, &c., were discovered within the
memory of history, which could not have happened if the
world had been eternal; and that there were visible marks
of ruin upon it which could not be effected by a flood only.
When I asked him how this earth could have been re-
peopled if ever it had undergone the same fate it was
threatened with hereafter by the comet of 1680, he answered, that required the power of a Creator. He said he
took all the planets to be composed of the same matter
with this earth, viz., earth, water, stones, &c., but variously
concocted. I asked him why he would not publish his
conjectures as conjectures, and instanced that Kepler had
communicated his; and though he had not gone near so
far as Kepler, yet Kepler's guesses were so just and happy,
that they had been proved and demonstrated by him. His
answer was, 'I do not deal in conjectures.' But upon my
talking to him about the four observations that had been
made of the comet of 1680, at five hundred and seventy-
four years' distance, and asking him the particular times,
he opened his *Principia*, which laid on the table, and
showed me there the particular periods, viz., 1st, the Julium
Sidus—in the time of Justinian—in 1106—in 1680.

"And I observing that he said there of that comet,
'Incidet in corpus solis,' and in the next paragraph adds,
'Stellæ fixæ refici possunt,' told him I thought he owned
there what he had been talking about, viz., that the comet

would drop into the sun, and that fixed stars were
recruited and replenished by comets when they dropt into
them; and consequently, that the sun would be recruited
too; and asked him why he would not own as freely what
he thought of the sun as well as what he thought of the
fixed stars. He said, 'That concerned us more;' and
laughing, added, 'that he had said enough for people to
know his meaning.'"

The preceding paper, with the title prefixed to it, was
first published by Mr. Turnor in his *Collections, &c.*, p. 172.
It was found among the Portsmouth manuscripts, in the
handwriting of Mr. Conduitt.

No. IV.

DEATH OF NEWTON'S MOTHER.

THE course of this work has been so continuously occupied with the discoveries and scientific and literary labours
of Sir Isaac Newton, that a convenient place has not
occurred for mentioning the death of his mother. It
took place in 1689, at the time Newton was one of the
representatives of the University of Cambridge in the
Convention Parliament. She went to Stamford, in Lincolnshire, to nurse her son, Benjamin Smith, Newton's
half-brother, who was taken ill with a malignant fever
there; and was in consequence seized with the same complaint herself. Newton left his duties and his studies to
watch at her bedside, sitting up whole nights with her,
administering her medicines, and preparing her blisters
with the most filial care and solicitude. She sank notwithstanding under the disease, and her remains were
carried to Colsterworth and deposited in the north aisle of
the church, where the family had generally been interred.

INAUGURATION

OF

THE STATUE OF SIR ISAAC NEWTON.

AFTER a lapse of nearly two centuries, a statue has been erected in honour of this great philosopher at Grantham, the locality of his boyhood, and where were sown the germs of that mathematical lore which rendered his name famous in his own country and in his own day, as well as in all lands, and for all ages.

The inauguration took place on Tuesday, the 21st of September, 1858. There was a procession from the Grammar School, where Newton was educated, to St. Peter's-hill. The procession consisted of the mayor and other municipal authorities, the clergy, headed by the Bishop of the diocese, Lord Brougham, the Master of Trinity College, Cambridge, the Master of the Mint, the Committee, and the gentlemen attending by invitation. There were also the boys from the Grammar School; the head boy carrying the "Principia;" the second, the reflecting telescope, made by Newton; and the third, Newton's prism, also made by him. One feature of the celebration — by no means the least remarkable, or the least gratifying — was, that the principal instrument of doing honour to England's greatest son in mathematical and astronomical science, was the venerable nobleman whose whole life had been devoted to the diffusion of general-knowledge, particularly of that science of which Newton was the great discoverer, as well as of all that is calculated to promote the cause of humanity and the amelioration of the race.

A semi-circular platform, erected for the accommodation of the spectators, was occupied by a numerous assemblage from the district, as well as from the more distant places.

The statue, which was covered with drapery, is cast in bronze, and is the work of W. Theed, Esq., whose established reputation as an artist of the highest class rendered him eminently qualified to execute the important task intrusted to him. He had been a student in Rome for twenty-two years, where he had the constant advice of the famous Thorwaldsen, the Danish sculptor, as also that of our justly-celebrated countryman Gibson. Nor have the expectations, which the fame of Mr. Theed was calculated to raise, been in any way disappointed; for it is generally admitted that he has produced a work which is highly creditable to himself, and a testimonial not unworthy even of the great philosopher in whose honour it is erected.

The statue is placed on a vacant piece of ground at the south end of High Street, lately known as Wood Hill, but to which the original name of St. Peter's Hill has been restored, and faces the west, looking along the road which Sir Isaac must have passed whenever he came to Grantham. It represents the philosopher in the costume of the period, and in the gown of a master of arts, in the act of lecturing, and his right hand is pointing to a celebrated diagram, taken from the " Principia," drawn upon the scroll in his left hand. The likeness is from the well-known mask of Sir Isaac's face, taken after death, and from the portrait bust by Ronbiliac. The figure is twelve feet high, and weighs upwards of two tons, about one half of which quantity was presented in the shape of old cannon by her Majesty's Government. It was cast at the foundry of Messrs. Robinson and Cottam, of Pimlico, and does them the highest credit. The pedestal on which the figure stands is from a design by Mr. Theed; it is fourteen feet high, and was cut from the celebrated marble quarries near Holyhead.   The total height of the pedestal and figure together is twenty-six feet.

The expense of the monument was raised by public subscription, amounting to £1,630, of which sum her Majesty and the Prince Consort contributed £100, and Grantham and the neighbourhood £600.

On a raised platform in front of the statue were placed two state chairs, of very beautiful design, the property, we believe, of the corporation of Grantham, and here were seated the bishop of the diocese and the mayor of the town.   In front of these seats was a third chair—an arm-chair—to which Lord Brougham was taken by the mayor.   It was such a chair as one sees occasionally in a country mansion—more rarely in a farm-house—straight in the back, with carved arms and legs, slightly carved all over, and a horse-hair cushion, covered with—what had been in its time, when it came from the hands of its maker—crimson moreen, but now faded and discoloured by age and wear.   The wear had, in fact, been so hard, as to reveal along the whole length of its front the material with which it had been stuffed.   It was the chair used by Newton himself nearly two hundred years ago, and the moreen was thus worn away while the original illustrious owner was composing the " Principia."

As soon as his lordship had taken his seat in this honoured relic, a signal was given by the mayor, and the covering which concealed the statue immediately fell to the ground.   Its exposure to view was received with an enthusiastic cheer by the assembled thousands.

Lord Brougham then at once rose and delivered the following oration :—

To record the names, and preserve the memory of those

whose great achievements in science, in arts, or in arms
have conferred benefits and lustre upon our kind, has in
all ages been regarded as a duty and felt as a gratification
by wise and reflecting men. The desire of inspiring an
ambition to emulate such examples generally mingles itself
with these sentiments; but they cease not to operate even
in the rare instances of transcendent merit, where matchless genius excludes all possibility of imitation, and nothing
remains but wonder in those who contemplate its triumphs
at a distance that forbids all attempts to approach. We
are this day assembled to commemorate him of whom the
consent of nations has declared that he is chargeable with
nothing like a follower's exaggeration or local partiality,
who pronounces the name of NEWTON as that of the
greatest genius ever bestowed by the bounty of Providence
for instructing mankind on the frame of the universe, and
the laws by which it is governed.

> Qui genus humanum ingenio superavit, et omnes
> Restinxit, stellas exortus uti ætherius sol.—(*Lucretius.*)

> In genius who surpassed mankind as far
> As does the mid-day sun the midnight star.—(*Dryden.*)

But though scaling these lofty heights be hopeless, yet
is there some use and much gratification in contemplating
by what steps he ascended. Tracing his course of action
may help others to gain the lower eminences lying within
their reach; while admiration excited and curiosity satisfied
are frames of mind both wholesome and pleasing. Nothing
new, it is true, can be given in narrative, hardly anything
in reflection, less still perhaps in comment or illustration;
but it is well to assemble in one view various parts of the
vast subject, with the surrounding circumstances, whether
accidental or intrinsic, and to mark in passing the mis-
conceptions raised by individual ignorance, or national

prejudice, which the historian of science occasionally finds crossing his path.

The remark is common, and is obvious, that the genius of Newton did not manifest itself at a very early age. His faculties were not, like those of some great and many ordinary individuals, precociously developed. Among the former, Clairaut stands pre-eminent, who at nineteen years of age presented to the Royal Academy a memoir of great originality upon a difficult subject in the higher geometry; and, at eighteen, published his great work on curves of double curvature, composed during the two preceding years. Pascal, too, at sixteen, wrote an excellent treatise on conic sections. That Newton cannot be ranked in this respect with those extraordinary persons, is owing to the accidents which prevented him from entering upon mathematical study before his eighteenth year; and then a much greater marvel was wrought than even the Clairauts and the Pascals displayed. His earliest history is involved in some obscurity; and the most celebrated of men has in this particular been compared to the most celebrated of rivers, as if the course of both in its feebler state had been concealed from mortal eyes. We have it, however, well ascertained that within four years, between the age of eighteen and twenty-two, he had begun to study mathematical science, and had taken his place among its greatest masters, learnt for the first time the elements of geometry and analysis, and discovered a calculus which entirely changed the face of the science, effecting a complete revolution in that and in every branch of philosophy connected with it. Before 1661 he had not read Euclid; in 1665 he had committed to writing the method of fluxions. At twenty-five years of age he had discovered the law of gravitation, and laid the foundations of celestial dynamics, the science created by him. Before ten years had elapsed, he added to his discoveries that of the fundamental pro-

perties of light.—So brilliant a course of discovery, in so
short a time, changing and reconstructing analytical,
astronomical, and optical science, almost defies belief.
The statement could only be deemed possible by an appeal
to the incontestable evidence that proves it strictly true.

By a rare felicity these doctrines gained the universal
assent of mankind as soon as they were clearly understood;
and their originality has never been seriously called in
question.   Some doubts having been raised respecting his
inventing the calculus, doubts raised in consequence of his
so long withholding the publication of his method, no
sooner was inquiry instituted than the evidence produced
proved so decisive, that all men in all countries acknowledged him to have been by several years the earliest inventor, and Leibnitz, at the utmost, the first publisher;
the only questions raised being, first, whether or not he
had borrowed from Newton, and next, whether, as second
inventor, he could have any merit at all; both which
questions have long since been decided in favour of
Leibnitz.

But undeniable though it be that Newton made the
great steps of this progress, and made them without any
anticipation or participation by others, it is equally certain
that there had been approaches in former times by preceding philosophers to the same discoveries.   Cavalieri by
his Geometry of Indivisibles, (1635,) Roberval by his
Method of Tangents, (1637,) had both given solutions
which Descartes could not attempt; and it is remarkable
that Cavalieri regarded curves as polygons, surfaces as
composed of lines, whilst Roberval viewed geometrical
quantities as generated by motion; so that the one approached to the differential calculus, the other to fluxions;
and Fermat, in the interval between them, comes still
nearer the great discovery by his determination of maxima
and minima, and his drawing of tangents.   More recently

Schooten had made public similar methods invented by Hudde; and, what is material, treating the subject algebraically, while those just now mentioned had rather dealt with it geometrically. It is thus easy to perceive how near an approach had been made to the calculus before the great event of its final discovery.

There had in like manner been approaches made to the law of gravitation, and the dynamical system of the universe. Galileo's important propositions on motion, especially on curvilinear motion, and Kepler's laws upon the elliptical form of the planetary orbits, the proportion of the areas to the times, and of the periodic times to the mean distances, and Huygens's theorems on centrifugal force, had been followed by still nearer approaches to the doctrine of attraction. Borelli had distinctly ascribed the motion of satellites to their being drawn towards the principal planets, and thus prevented from being carried off by the centrifugal force.

Even the composition of white light, and the different action of bodies upon its component parts, had been vaguely conjectured by Anth. de Dominis, Archbishop of Spalatro, at the beginning, and more precisely in the middle of the seventeenth century by Marcus, (Kronland of Prague,) unknown to Newton, who only refers to the Archbishop's work; while the treatise of Huygens on light, Grimaldi's observations on colours by inflexion as well as on the elongation of the image in the prismatic spectrum, had been brought to his attention, although much less near to his own great discovery than Marcus's experiment.

But all this only shows that the discoveries of Newton, great and rapid as were the steps by which they advanced our knowledge, yet obeyed the law of continuity, or rather of gradual progress, which governs all human approaches towards perfection. The limited nature of man's faculties precludes the possibility of his ever reaching at once the

utmost excellence of which they are capable.   Survey the
whole circle of the sciences, and trace the history of our
progress in each, you find this to be the universal rule.   In
chemical philosophy the dreams of the alchemists prepared
the way for the more rational though erroneous theory of
Stahl ; and it was by repeated improvements that his errors,
so long prevalent, were at length exploded, giving place
to the sound doctrine which is now established.   The great
discoveries of Black and Priestley on heat and aëriform
fluids had been preceded by the happy conjectures of
Newton, and the experiments of others.   Nay, Voltaire had
well-nigh discovered both the absorption of heat, the constitution of the atmosphere, and the oxidation of metals,
and by a few more trials might have ascertained it.

Cuvier had been preceded by inquirers who took sound
views of fossil osteology ; among whom the truly original
genius of Hunter fills the foremost place.   The inductive
system of Bacon had been, at least in its practice, known
to his predecessors.   Observations and even experiments
were not unknown to the ancient philosophers, though
mingled with gross errors ; in early times, almost in the
dark ages, experimental inquiries had been carried on with
success by Friar Bacon, and that method actually recommended in a treatise, as it was two centuries later by
Lionardo da Vinci ; and at the latter end of the next century Gilbert examined the whole subject of magnetic action
entirely by experiments.   So that Lord Bacon's claim to
be regarded as the father of modern philosophy rests upon
the important, the invaluable step of reducing to system
the method of investigation adopted by those eminent men,
generalizing it, and extending its application to all matters
of contingent truth, exploding the errors, the absurd
dogmas, and fantastic subtleties of the ancient schools,
above all confining the subject of our inquiry, and the

manner of conducting it, within the limits which our
faculties prescribe.

Nor is this great law of gradual progress confined to
the physical sciences; in the moral it equally governs.
Before the foundations of political economy were laid by
Hume and Smith, a great step had been made by the
French philosophers, disciples of Quesnay; but a nearer
approach to sound principles had signalized the labours of
Gournay, and those labours had been shared and his doctrines patronized by Turgôt when Chief Minister. Again,
in constitutional policy, see by what slow degrees, from its
first rude elements,—the attendance of feudal tenants at
their lord's court, and the summons of burghers to grant
supplies of money,—the great discovery of modern times
in the science of practical politics has been effected, the
representative scheme, which enables States of any extent
to enjoy popular government, and allows mixed monarchy
to be established, combining freedom with order; a plan
pronounced by the statesmen and writers of antiquity to
be of hardly possible formation, and wholly impossible
continuance. The globe itself, as well as the science of
its inhabitants, has been explored according to the law
which forbids a sudden and rapid leaping forward, and
decrees that each successive step, prepared by the last,
shall facilitate the next. Even Columbus followed several
successful discoverers on a smaller scale; and is by some
believed to have had, unknown to him, a predecessor in
the great exploit by which he pierced the night of ages,
and unfolded a new world to the eyes of the old.

The arts afford no exception to the general law. De-
mosthenes had eminent forerunners, Pericles the last of
them. Homer must have had predecessors of great merit,
though doubtless as far surpassed by him, as Fra Barto-
lomeo and Pietro Perugino were by Michael Angelo and
Raphael. Dante owed much to Virgil; he may be allowed

to have owed, through his Latin Mentor, not a little to the old Grecian; and Milton had both the orators and the poets of the ancient world for his predecessors and his masters. The art of war itself is no exception to the rule. The plan of bringing an overpowering force to bear on a given point had been tried occasionally before Frederick II. reduced it to a system; and the Wellingtons and Napoleons of our own day made it the foundation of their strategy, as it had also been previously the mainspring of our naval tactics.

It has oftentimes been held that the invention of logarithms stands alone in the history of science, as having been preceded by no step leading towards the discovery. There is, however, great inaccuracy in this statement; for not only was the doctrine of infinitesimals familiar to its illustrious author, and the relation of geometrical to arithmetical series well known; but he had himself struck out several methods of great ingenuity and utility, (as that known by the name of *Napier's Bones*,)—methods that are now forgotten, eclipsed as they were by the consummation which has immortalized his name. So the inventive powers of Watt, preceded as he was by Worcester and Newcomen, but more materially by Caus and Papin, had been exercised on some admirable contrivances, now forgotten, before he made the step which created the engine anew, not only the parallel motion, possibly a corollary to the proposition on circular motion in the Principia, but the Separate Condensation, and above all the Governor, perhaps the most exquisite of mechanical inventions; and now we have those here present who apply the like principle to the diffusion of knowledge, aware, as they must be, that its expansion has the same happy effect naturally of preventing mischief from its excess, which the skill of the great mechanist gave artificially to steam, thus rendering his engine as safe as it is powerful.

The grand difference, then, between one discovery or invention and another is in degree rather than in kind; the degree in which a person, while he outstrips those whom he comes after, also lives as it were before his age. Nor can any doubt exist that in this respect Newton stands at the head of all who have extended the bounds of knowledge. The sciences of dynamics and of optics are especially to be regarded in this point of view; but the former in particular; and the completeness of the system which he unfolded, its having been at the first elaborated and given in perfection, its having, however new, stood the test of time, and survived, nay, gained by, the most rigorous scrutiny, can be predicated of this system alone, at least in the same degree. That the calculus, and those parts of dynamics which are purely mathematical, should thus endure for ever, is a matter of course. But his system of the universe rests partly upon contingent truths, and might have yielded to new experiments, and more extended observation. Nay, at times it has been thought to fail, and further investigation was deemed requisite to ascertain if any error had been introduced; if any circumstance had escaped the notice of the great founder. The most memorable instance of this kind is the discrepancy supposed to have been found between the theory and the fact in the motion of the lunar apsides, which about the middle of the last century occupied the three first analysts of the age. The error was discovered by themselves to have been their own in the process of their investigation; and this, like all the other doubts that were ever momentarily entertained, only led in each instance to new and more brilliant triumphs of the system.

The prodigious superiority in this cardinal point of the Newtonian to other discoveries appears manifest upon examining almost any of the chapters in the history of science. Successive improvements have, by extending our views, constantly displaced the systems that appeared firmly established. To take a familiar instance, how little remains of Lavoisier's doctrine of combustion and acidification, except the negative positions, the subversion of the system of Stahl! The substance having most eminently the properties of an acid (chlorine), is found to have no oxygen at all, while many substances abounding in oxygen, including alkalis themselves, have no acid property whatever; and without the access of oxygenous or of any other gas, heat and flame are produced in excess. The doctrines of free trade had not long been promulgated by Smith, before Bentham demonstrated that his exception of usury was groundless; and his theory has been repeatedly proved erroneous on colonial establishments, as his exception to it on the navigation laws; while the imperfection of his views on the nature of rent is undeniable, as well as on the principle of population. In these and such instances as these, it would not be easy to find in the original doctrines the means of correcting subsequent errors, or the germs of extended discovery. But even if philosophers finally adopt the undulatory theory of light instead of the atomic, it must be borne in mind that Newton gave the first elements of it by the well-known proposition in Section VIII. of the second Book of the Principia; the scholium to that section also indicating his expectation that it would be applied to optical science; while M. Biot has shown how the doctrine of fits of reflexion and transmission tallies with polarization, if not with undulation also.

But the most marvellous attribute of Newton's discoveries is that in which they stand out prominent among all the other feats of scientific research, stamped with the peculiarity of his intellectual character; they were (for their great author lived before his age) anticipating in part what was long after wholly accomplished; and thus unfolding some things which at the time could be but imperfoctly, others not at all, comprehended; and not rarely pointing out the path and affording the means of treading it to the ascertainment of truths then veiled in darkness. He not only enlarged the actual dominion of knowledge, penetrating to regions never before explored, and taking with a firm hand undisputed possession; but he showed how the bounds of the visible horizon might be yet further extended, and enabled his successors to occupy what he could only descry; as the illustrious discoverer of the new world made the inhabitants of the old cast their eyes over lands and seas far distant from those he had traversed; lands and seas of which they could form to themselves no conception, any more than they had been able to comprehend the course by which he led them on his grand enterprise. In this achievement, and in the qualities which alone made it possible—inexhaustible fertility of resources, patience unsubdued, close meditation that would suffer no distraction, steady determination to pursue paths that seemed all but hopeless, and unflinching courage to declare the truths they led to, how far soever removed from ordinary apprehension,—in these characteristics of high and original genius we may be permitted to compare the career of those great men. But Columbus did not invent the mariner's compass, as Newton did the instrument which guided his course and enabled him to make his discoveries, and his successors to extend them by closely following his directions in using it. Nor did the compass suffice to the great navigator without making any observations—though he dared to steer without a chart; while it is certain that by the philosopher's instrument his discoveries were extended over the whole system of the universe, determining the masses, the forms, and the motions of all its parts, by the mere inspection of abstract calculations and formulae analytically deduced.

The two great improvements in this instrument which

have been made, the calculus of variations by Euler and Lagrange, the method of partial differences by D'Alembert, we have every reason to believe, were known, at least in part, to Newton himself. His having solved an isoperimetrical problem (finding the line whose revolution forms the solid of least resistance) shows clearly that he must have made the co-ordinates of the generating curve vary, and his construction agrees exactly with the equation given by that calculus. That he must have tried the process of integrating by parts in attempting to generalize the inverse problem of central forces before he had recourse to the geometrical approximation which he has given, and also when he sought the means of ascertaining the comet's path, (which he has termed by far the most difficult of problems,) is eminently probable, when we consider how naturally that method flows from the ordinary process for differentiating compound quantities by supposing each variable in succession constant; in short, differentiating by parts. As to the calculus of variations having substantially been known to him, no doubt can be entertained.

Again, in estimating the ellipticity of the earth, he proceeded upon the assumption of a proposition of which he gave no demonstration, (any more than he had done of the isoperimetrical problem,) that the ratio of the centrifugal force to gravitation determines the ellipticity. Half a century later, that which no one before knew to be true, which many probably considered to be erroneous, was examined by one of his most distinguished followers, Maclaurin, and demonstrated most satisfactorily to be true.

Newton had not failed to perceive the necessary effects of gravitation in producing other phenomena beside the regular motion of the planets and their satellites, in their course round their several centres of attraction. One of

these phenomena, wholly unsuspected before the discovery
of the general law, is the alternate movement to and fro
of the earth's axis, in consequence of the solar (and also
of the lunar) attraction combined with the earth's motion.
This libration, or nutation, distinctly announced by him
as the result of the theory, was not found by actual
observation to exist till sixty years and upwards had
elapsed, when Bradley proved the fact.

The great discoveries which have been made by
Lagrange and Laplace upon the results of disturbing
forces, have established the law of periodical variation of
orbits, which secures the stability of the system by pre-
scribing a maximum and a minimum amount of deviation ;
and this is not a contingent but a necessary truth, by
rigorous demonstration, the inevitable result of undoubted
data in point of fact, the eccentricities of the orbits, the
directions of the motions, and the movement in one plane
of a certain position. That wonderful proposition of
Newton, which with its corollaries may be said to give the
whole doctrine of disturbing forces, has been little more
than applied and extended by the labours of succeeding
geometricians. Indeed, Laplace, struck with wonder at
one of his comprehensive general statements on disturbing
forces in another proposition, has not hesitated to assert,
that it contains the germ of Lagrange's celebrated inquiry,
exactly a century after the Principia was given to the
world.

The wonderful powers of generalization, combined with
the boldness of never shrinking from a conclusion that
seemed the legitimate result of his investigations, how new
and even startling soever it might appear, was strikingly
shown in that memorable inference which he drew from
optical phenomena, that the diamond is "an unctuous
substance coagulated;" subsequent discoveries having
proved both that such substances are carbonaceous, and

that the diamond is crystallized carbon; and the foundations of mechanical chemistry were laid by him with the boldest induction and most felicitous anticipations of what has since been effected. The solution of the inverse problem of disturbing forces has led Le Verrier and Adams to the discovery of a new planet, merely by deductions from the manner in which the motions of an old one are affected, and its orbit has been so calculated that observers could find it; nay, its disc, as measured by them, only varies one twelve-hundredth part of a degree from the amount given by the theory. Moreover, when Newton gave his estimate of the earth's density, he wrote a century before Maskelyne, by measuring the force of gravitation in the Scotch mountains, 1772, gave the proportion to water as 4·716 to 1; and many years after by experiments with mechanical apparatus Cavendish, 1798, corrected this to 5·48; and Baily more recently, 1842, to 5·66, Newton having given the proportion as between 5 and 6 times. In these instances he only showed the way, and anticipated the result of future inquiry by his followers. But the oblate figure of the earth affords an example of the same kind, with this difference, that here he has himself perfected the discovery, and nearly completed the demonstration. From the mutual gravitation of the particles which form its mass, combined with their motion round its axis, he deduced the proposition that it must be flattened at the poles; and he calculated the proportion of its polar to its equatorial diameter. By a most refined process he gave this proportion upon the supposition of the mass being homogeneous. That the proportion is different in consequence of the mass being heterogeneous does not in the least affect the soundness of his conclusion. Accurate measurements of a degree of latitude in the equatorial and polar regions, with experiments on the force of gravitation in those regions, by the different lengths of

a pendulum vibrating seconds, have shown that the excess
of the equatorial diameter is about eleven miles less than
he had deduced it from the theory; and thus that the
globe is not homogeneous, but on the assumption of a
fluid mass, the ground of his hydrostatical investigation,
his proportion of 229 to 230 remains unshaken; and is
precisely the one adopted and reasoned from by Laplace,
after all the improvements and all the discoveries of later
times.   Surely at this we may well stand amazed, if not
awe-struck.   A century of study, of improvement, of
discovery, has passed away; and we find Laplace, master
of all the new resources of the calculus, and occupying
the heights to which the labours of Euler, Clairaut,
D'Alembert, and Lagrange have enabled us to ascend,
adopting the Newtonian fraction of one two-hundred-and
thirtieth, as the accurate solution of this speculative
problem.   New admeasurements have been undertaken
upon a vast scale, patronized by the munificence of rival
governments; new experiments have been performed with
improved apparatus of exquisite delicacy; new observations
have been accumulated, with glasses far exceeding any
powers possessed by the resources of optics in the days of
him to whom the science of optics, as well as dynamics,
owes its origin; the theory and the fact have thus been
compared and reconciled together in more perfect harmony;
but that theory has remained unimproved, and the great
principle of gravitation, with its most sublime results,
now stands in the attitude, and of the dimensions, and
with the symmetry, which both the law and its application
received at once from the mighty hand of its immortal
author.

But the contemplation of Newton's discoveries raises
other feelings than wonder at his matchless genius.   The
light with which it shines is not more dazzling than useful.
The difficulties of his course, and his expedients, alike

copious and refined, for surmounting them, exercise the
faculties of the wise while commanding their admiration;
but the results of his investigations, often abstruse, are
truths so grand and comprehensive, yet so plain, that they
both captivate and instruct the simple. The gratitude,
too, which they inspire, and the veneration with which
they encircle his name, far from tending to obstruct
future improvement, only proclaim his disciples the zealous
because rational followers of one whose example both
encouraged and enabled his successors to make further
progress. How unlike the blind devotion to a master
which for so many ages of the modern world paralysed
the energies of the human mind!—

> " Had we still paid that homage to a name
> Which only God and nature justly claim,
> The western seas had been our utmost bound,
> And poets still might dream the sun was drown'd,
> And all the stars that shine in southern skies
> Had been admired by none but savage eyes."

Nor let it be imagined that the feelings of wonder
excited by contemplating the achievements of this great
man are in any degree whatever the result of national
partiality, and confined to the country which glories in
having given him birth. The language which expresses
her veneration is equalled, perhaps exceeded, by that in
which other nations give utterance to theirs: not merely
by the general voice, but by the well-considered and well-
informed judgment of the masters of science. Leibnitz,
when asked at the royal table in Berlin his opinion of
Newton, said that, "taking mathematicians from the
beginning of the world to the time when Newton lived,
what he had done was much the better half." "The
Principia will ever remain a monument of the profound
genius which revealed to us the greatest law of the
universe," are the words of Laplace. "That work stands

pre-eminent above all the other productions of the human mind. The discovery of that simple and general law by the greatness and the variety of the objects which it embraces, confers honour upon the intellect of man." Lagrange, we are told by Delambre, was wont to describe Newton as the greatest genius that ever existed, but to add how fortunate he was also, "because there can only once be found a system of the universe to establish." "Never," says the father of the Institute of France, one filling a high place among the most eminent of its members,—"Never," says M. Biot, "was the supremacy of intellect so justly established and so fully confessed. In mathematical and in experimental science without an equal and without an example; combining the genius for both in its highest degree." The Principia he terms the greatest work ever produced by the mind of man, adding, in the words of Halley, that a nearer approach to the divine nature has not been permitted to mortals. "In first giving to the world Newton's method of fluxions," says Fontenelle, "Leibnitz did like Prometheus; he stole fire from heaven to teach men the secret." "Does Newton," L'Hôpital asked, "sleep and wake like other men? I figure him to myself as of a celestial kind, wholly severed from mortality."

To so renowned a benefactor of the world, thus exalted to the loftiest place by the common consent of all men, one whose life without the intermission of an hour was passed in the search after truths the most important, and at whose hands the human race had only received good, never evil, those nations have raised no memorial which erected statues to the tyrants and conquerors, the scourges of mankind, whose lives were passed not in the pursuit of truth but the practice of falsehood, across whose lips if truth ever chanced to stray towards some selfish end, it surely failed to obtain belief; who, to slake their insane

thirst of power, or of pre-eminence, trampled on all the
rights, and squandered the blood, of their fellow-creatures;
whose course, like the lightning, blasted while it dazzled;
and who, reversing the noble regret of the Roman Emperor, deemed the day lost that saw the sun go down upon
their forbearance, no victim deceived, or betrayed, or op-
pressed. That the worshippers of such pestilent genius
should consecrate no outward symbol of the admiration
they freely confessed to the memory of the most illustrious
of men, is not matter of wonder. But that his own countrymen, justly proud of having lived in his time, should have
left this duty to their successors, after a century and a
half of professed veneration and lip homage, may well be
deemed strange. The inscription upon the cathedral
masterpiece of his celebrated friend's architecture, may
possibly be applied in defence of this neglect:—"If you
seek for a monument, look around." If you seek for a
monument, lift up your eyes to the heavens which show
forth his fame. Nor, when we recollect the Greek orator's
exclamation, "The whole earth is the monument of illustrious men," can we stop short of declaring that the whole
universe is Newton's. Yet in raising the statue which
preserves his likeness near the place of his birth, on the
spot where his prodigious faculties were unfolded and
trained, we at once gratify our honest pride as citizens of
the same state, and humbly testify our grateful sense of
the Divine goodness which deigned to bestow upon our
race one so marvellously gifted to comprehend the works
of Infinite Wisdom, and so piously resolved to make all
his study of them the source of religious contemplations,
both philosophic and sublime.

FINIS.

WILLIAM TEGG & CO.'S

LIST OF

PUBLICATIONS.

, *Booksellers can be supplied with this Catalogue, with their Name and Address, to the extent of 150 Copies.*

LONDON:
12, PANCRAS LANE, CHEAPSIDE.

Two Vols., Crown 8vo.

THINGS NEW & OLD:

OR,

A STOREHOUSE OF SIMILES,
SENTENCES, ALLEGORIES, APOPHTHEGMS,
ADAGES, APOLOGUES, Etc.

WITH THEIR SEVERAL APPLICATIONS.

*Collected and Observed from the Writings and Sayings of
the Learned in all Ages to this Present.*

By JOHN SPENCER,

A LOVER OF LEARNING AND LEARNED MEN,

WITH

A PREFACE BY THE REV. THOMAS FULLER, D.D.

Price 12s. 6d.

Crown 8vo.

THE

LIFE OF PETER THE GREAT,

CZAR OF RUSSIA.

By J. BARROW, F.R.S.,
Author of " The Mutiny of the Bounty."
ILLUSTRATED.

Price 6s.

Demy 8vo.

MILTON'S (JOHN)

POETICAL WORKS,

EDITED BY SIR E. BRYDGES, BART.

Illustrated by J. M. W. TURNER, R.A. *New Edition carefully corrected.*
Price 15s.; morroco, 24s.

8vo.

THE KORAN.

TRANSLATED INTO ENGLISH IMMEDIATELY FROM THE
ORIGINAL ARABIC, WITH EXPLANATORY NOTES.

BY GEORGE SALE.

A NEW EDITION, WITH VARIOUS READINGS AND NOTES
(SAVERY'S VERSION), WITH PLAN AND VIEW OF
BEIH ALLAH.

Price 8s. 6d.

Small Crown 8vo.

SCIENTIFIC TERMS

(THE DICTIONARY OF).

Explanatory of all the Terms used in the Arts,

Sciences, &c., &c.

By W. M. BUCHANAN.

A NEW EDITION.

Price 6s.

LONDON:

WILLIAM TEGG & CO., PANCRAS LANE, CHEAPSIDE.